How to Cultivate a World-class Safety Culture

Actively Engaging Employees Using the Five Pillars of Safety

MICHAEL SAUJANI

This book is dedicated to my late parents. My dad was denied his final year of education in high school when his elder brother lifted him from school and took him to Uganda to run a *duka* or shop. As a result, he always wanted his children to be educated and literally forced me to stay away from his duka and continue to improve my knowledge through studying. His next passion was adventure, exploring new things. So when I retired, thanks to him, I ventured into writing this book—something I never imagined I could do. It has been a good adventure and a productive experience.

My mom, on the other hand, was light-hearted and loving. As a result, you will notice light-hearted stories in this otherwise serious book.

Again, thanks dad and mom. This book would not have been possible without your blessings.

Contents

APPENDIX

Foreword

My father has always been concerned about safety—even as a Dad. Whether it was making sure we had flashlights on our family camping trips, reminding us to turn off the stove when we first moved into our own apartments or making sure we were up to date on our car insurance payments, my father was always concerned about safety. He created a culture in our family that valued safety. As my sister and I grew up, we brought these values to our own families and organizations.

This book is my father's labor of love. For years, in every leadership role my father had, he would come home at night and write notes about the lessons and stories of safety he learned that day or week in the hopes that he could inspire others. He always wanted to use his passion for safety to teach others. My dad has always been a gifted teacher. In this book, he takes the lessons he learned as a mechanical and safety engineer for the past fifty years and shares them with you so that you and your organization may also build a world-class safety culture.

The insights in this book are invaluable. It is a collection of fifty years of knowledge across countries and industries. The success he has had in developing environments that promote safety can now be shared with other employers who are similarly interested in building systems and processes that create healthy environments. The Five Pillars of Safety that he has developed are accessible to all who want to build a world-class safety culture. Any person or organization can create a culture that values safety, which is the guiding principle my father hopes to leave you with. I hope you walk away from this book as I did: with a deeper appreciation for how building a world-class safety culture can make a lasting impact on an organization.

Reshma Saujani

Preface

I have been in the safety and loss prevention field for more than forty-five years. First as a factories inspector for the Department of Labor in the Government of Uganda, where I inspected cotton ginneries, sugar factories, and textile mills. Then I moved to the USA, where over the years I have worked as safety engineer for American Mutual Insurance Company, loss-prevention consultant for Hartford Insurance, and loss-control supervisor for Hanover Insurance Company. Later I worked as corporate safety director for Fort Dearborn Company (FDC), a large label printing operation. I was able to use my technical education and communication skills, developed while working for various insurance companies, at the printing company and change its culture from "OSHA's Most Wanted" to the "Best in Class."

By education I am a mechanical engineer, having passed my Bachelor of Engineering degree within the top twenty of my university class and earning the recognition of First Class with Distinction. I took several graduate level courses at Illinois Institute of Technology in Chicago and later engulfed myself in insurance and safety-related education and training as I found the field both emotionally and financially rewarding. In the process, I acquired the designations of Associate in Loss Control Management (ALCM), Associate in Risk Management (ARM), Associate in Claims Management (ACM), and

Certified Property and Casualty Underwriters (CPCU). During my first year at American Mutual Insurance, while taking classes at IIT, I sat for and passed the examination securing the Certified Safety Profession (CSP) designation. I have spent a significant amount of time continuously improving my education and skills so that I can be an outstanding safety professional helping companies provide a safe place of employment for their employees.

I decided to write this book to give back to the safety profession that has given me so much. I want to share my experience with safety professionals so that they can use some of the lessons I learned to upgrade their career and implement safety initiatives for the betterment of the employees they are responsible for. In the following pages, I share the techniques I used in securing outstanding safety results at Fort Dearborn Company. I hope you will be able to do the same.

I would like to thank FDC for hiring me and allowing me to use my knowledge and skills in helping the company provide a safer environment for its employees. I had an outstanding thirteen years with the company and worked with all levels of management and employees to find ways to continuously improve the safety of its employees and customers. This book, which is dedicated to FDC's employees and management, lays down some of the techniques I used and the experiences I had during this tenure.

For safety professionals to be successful in developing an environment promoting safety, they need to have a purpose and vision. The purpose and vision should be, as it was mine, to develop and maintain a world-class safety culture. The book lays down an organized method using the Five Pillars of Safety to achieve this goal. It provides a road map for success. It describes examples and the techniques used to achieve outstanding safety results that you can easily apply and use to achieve similar results. As you apply these techniques, you will find ways to build upon them and reach an even higher level of awareness than I was able to.

I would like to give credit to my senior management for their whole-hearted support in achieving safety results and applying these techniques to protect company employees, its property, and customers.

Good luck!

MICHAEL SAUJANI,
CSP, CPCU, ALCM, ARM

How to Cultivate a
World-class Safety Culture

1

___

What It Means to Achieve World-class Safety

World-class safety culture (WCSC) is like beauty: it is difficult to describe, but you know it when you see it. In this book I will discuss and share how one company, Fort Dearborn Company (FDC), was able to develop world-class safety culture through innovation, hard work, and persistence.

In 2002 Fort Dearborn knew it had to take action to improve its safety performance. Two of the company's facilities had been included on OSHA's "Most Wanted" list and nearly all had higher than average lost-workday-incidence rates. To turn its safety program around, the company's executive committee, at the request of its insurance agency, authorized the hiring of a professional safety director. After a thorough search it decided to hire me because of my experience in the insurance industry as a safety professional. During the interview process I mentioned, "I do not know much about the printing industry." And the CEO responded, "You teach us safety and risk management and we will teach you the ins and outs of the printing industry." That is how our mutual relationship began, which lasted more than twelve years. During that time, I helped the company save insurance dollars and provide a safe environment for its employees and customers by developing a world-class safety culture. What is culture?

At a presentation in New Delhi called "Roadmap to World-class Safety," J. Murali defined safety culture as "the attitude, beliefs, perceptions and values that employees share in relation to safety" in an organization. When safety professionals talk about world-class, they generally mean best of the best, best in the class, best in the world. In the arena of sports, a world-class athlete is one who rises to the top and is the best in the field. This accolade is typically based on a singular value such as the number of gold medals or championships won, number of baskets dunked, goals saved or scored, wickets taken or centuries made, etc. Since safety is a process, world-class safety culture does not have one definitive, singular value that can be recognized. The Robert W. Campbell Institute identified five principal findings when they analyzed several leading EHS Management Systems used by various companies applying for National Safety Council safety awards: the company's EHS was on par with business performance, they employed a system-based approach to EHS, they strived to carry out continuous improvement, their EHS aligned with the organization's strategies and values, and they actively promoted health and safety on and off the job.

Whenever we talk about world-class safety, we're really talking about how to build a culture of safety. Culture is defined as:

1. The beliefs, customs, arts, experience, attitudes, and religion of a particular group of people at a particular place or time.
2. A way of thinking, behaving or working that exists in a country, place or organization.
3. The set of shared attitudes, values, goals, and practices that characterizes an institution or organization.
4. It is a way of life: the behaviors, beliefs, values, symbols, and rituals that are passed down and communicated from generation to generation.

I remember the days in the 70s when I was a factory inspector in Uganda. My responsibility was to inspect all factories in the eastern region of the country and investigate all serious and fatal accidents. The factories in this region were mostly woodworking shops, cotton ginneries, textile mills, and jaggery mills. One day I received an accident report stating a nomadic person had placed his right foot on the spinning cotton gin just for curiosity. Not believing what I read, I decided to make a special trip to investigate the accident in the Acholi district. Apparently, the factory owner had hired the worker from a nearby village and asked him to push raw cotton inside the spinning wheel where the seeds would be separated, leaving fluffy cotton. It was a boring task and after a while the person's curiosity got the better of him. He pushed his leg into the gin just to feel the spinning wheel! He was seriously hurt and lucky he was not pulled inside the spinning wheel.

As a nomadic person, he had not seen many machines. He had no knowledge of how the spinning wheel works or what kind of power it has to separate the seeds from the cotton. The company manager thought it was a simple task and did not evaluate this person's beliefs or knowledge before assigning him to this potentially dangerous task. Safety culture was missing.

When working with people with cultural differences, it is important to identify these differences and find a common ground to build upon. In the above case, the manager was from England and the worker was from an African village with a completely different set of knowledge, beliefs, and values. It is therefore very important to study and understand the cultural differences and motivational factors when working globally to implement meaningful safety systems.

Businesses and individuals have their traits or cultures as well. The culture of a business can be understood as organizational norms, values, habits, behaviors, beliefs, and vision. It has been noted that in highly successful organizations, a significant number of their members have similar beliefs and values.

When I was first hired by FDC, it was a family-owned enterprise and privately held with multiple facilities manufacturing cut-and-stack, film, pressure-sensitive, and digital-printed labels for customers in the food, beverage, paint, and retail industries.

Cut-and-stack labels are made by feeding rolls of paper through a machine that cuts sheets of paper to the desired dimensions. Palletized sheets of paper are used to print the desired labels on a printing press. The printed labels are cut using cutting machines, then stacked and wrapped with cellophane plastic material, and shipped to customers via common carrier.

Shrink-sleeve labels are made from rolls of plastic that are rolled into place on a flexographic press. Each press has several printing stations. As the plastic material is fed through each printing station, ink is applied to make the labels. The rolls of labels are slit, packaged, and shipped to the customer.

I found the business culture of this family-owned organization to be caring and considerate. It provided Christmas holidays and over-time work prior to holidays so employees could buy extra gifts for their family members. This generosity created a bond between the employer and employees. It was a given thing whether or not the company was busy. Then everything changed; FDC was purchased by an equity firm when some of the third-generation family members wanted to cash out. The family-owned company culture changed in order to be consistent with the new hedge fund company's culture. Previously, the mission statement from the CEO of the family-owned company was:

> The personal safety and health of each associate of the company is of primary importance. Our philosophy for safety recognizes that personal injuries are preventable. No job is so important, so urgent, that we cannot take time to think about safety and to do our job safely. Tasks done safely are orderly and efficient.

When the hedge fund company took over, the mission statement changed to:

> To be the leading prime label supplier to the consumer goods marketplace by offering a comprehensive and innovative suite of label solutions to consistently meet our customers' dynamic packaging needs and business objectives while providing our associates opportunities to enrich themselves and their careers. We are committed to customer intimacy and providing customers with value, service, and the highest quality attainable.

Our core values changed to:

Customer Intimacy: Understand our customers' needs and challenges to support their growth and strategies

Financial Discipline: Manage the business to drive top and bottom line results to ensure long-term sustainable growth

Performance Driven: Reward associates who drive results and continual improvement

The new CEO oversaw the transition of the company from a family-owned enterprise to a portfolio company for a private equity firm. The CEO led the establishment of a new culture, one that was customer centric and focused on growth. In line with that focus, Fort Dearborn acquired and integrated its largest competitor, Renaissance Mark. Under the new CEO's leadership, the company increased sales by 91 percent and EBITDA by eight times in the last four years.

The new management felt they needed to change company culture in order to improve financial results and promote the new owners' business philosophy. Both of which to them were more important than emphasizing safety culture. The new CEO realized that cultures are either created organically or through deliberate and consistent planning and action. He had to take careful steps to manage and promote the new culture effectively. This is particularly true since much of Western society (USA, Australia, England, and some other countries) favors individualism. More people in the West have an individualistic nature and value independent thinking unlike Germany and Japan. So it is a little more difficult to get people here to coalesce together around a group culture.

A company's safety culture changes based on company needs and the leader's mission. It takes leaders time and effort to make these changes. As safety professionals, we need to understand new company culture and a leader's motivation so that we can guide the leadership in promoting employee behaviors that we believe are important to maintaining a world-class safety culture. At FDC I met with the CEO and discussed how safety makes good business sense. A disciplined and consistent safety approach would help him improve his core values of customer intimacy, financial discipline, and performance-driven metrics. I challenged him to keep safety on par with other operational functions and to regularly evaluate safety performance.

The CEO discussed these core values at every safety and business meeting. He started letting people go who did not believe in his core values and hiring people who did. Safety performance improved as well, since as Safety Director I tried my best to align safety initiatives with his core values. I started getting commitments and guidance from visible senior management. In turn, that increased employee and middle level participation in safety activities. The safety results improved further and management started talking about our safety initiatives at all levels, including board of directors meetings. Safety started to become contagious.

Organizational culture represents a company's character, heart, and soul. It determines company's operating systems. To change an organization culture to become world-class can require a lot of work and patience. One method that can help in changing organizational culture is by using Role Models. There are one or two people in an organization that are role models for most employees. We had a General Manager of one of the plants who was absolutely liked by almost all other GMs because of how he operated his plant. So when I had to implement a comprehensive LOTO program throughout the organization I started with his plant. He went beyond what I had expected and prepared specific LOTO procedure for each machine with pictures, diagrams and specific instructions. He placed the procedure in a pouch and hung it on each machine for easy access. At the GM meeting he explained and showed to people what he had done in his plant for employee safety. Next thing I found out that almost all plants had completed similar LOTO procedures within a month.

Unlike world-class safety culture, where good things happen to you when you do good, the "zero accident" culture is an attempt to reach zero accidents without doing everything a world-class safety culture company would do. In the process of avoiding accidents to achieve a zero accident rate, employees may take short cuts and hide injuries. This zero accident culture encourages borderline unethical behavior. When zero accident culture is a philosophical strategy and mandates the prevention of all accidents instead of a numerical goal, it impacts workers' behavior negatively. However, a company culture that promotes zero incidents, zero safety violations, and zero unsafe acts and conditions is a culture conducive to world-class safety.

The Five Pillars of Safety can be used to develop a WCSC. One manufacturing company—a large multi-location printing operation with plants in various states in U.S. and Canada—that I was associated with was able to use this belief in developing a world-class safety culture for the benefit of its employees, community, and customers. These five beliefs became the five safety pillars on which the company's safety platform was built. Although I primarily applied these changes in culture to a manufacturing environment, most of the principles and ideas can be applied to construction and service industries as well.

Before one can apply these principles, it is important to complete a gap analysis in order to determine where the organization is currently (current state) in relation to safety culture. Once that is known, a road map for success can be developed for the desired state.

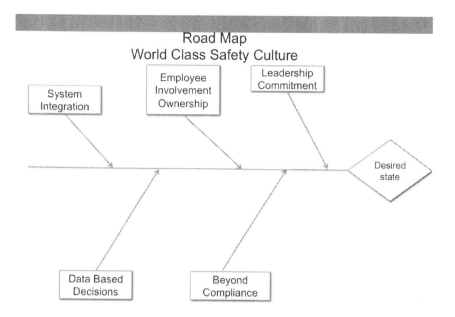

By definition, a road map is a detailed plan to guide progress toward a goal. It is a framework for guiding mid-level management and employees to understanding the actions and decisions needed so as to achieve the desired goal. Some of the elements can be misunderstood; it is the responsibility of the safety director to make sure everyone clearly understands each element of the road map. The road map shown above should include additional details based on the results of the gap analysis.

In a company that thrives on world-class safety, safety procedures should be seamless. Just like if one looks at the seam on a pair of pants, if you can't see the seam, people will be impressed with how well the pant was stitched. Only recently one of the company's new plant safety team leaders (every FDC plant has a safety team leader) said to me, "Mike, we do not have a safety program in place." He googled various sites on the web and, using his common sense, put together what he called a "robust safety program." He wanted to distribute this program to every plant and ask other leaders to start implementing the robust plan as the company had no safety program in place. Being a safety director responsible for the long term safety of a company's people and property, I said no. He did not like that. He meant well, but he had no safety education or background. This made me realize there is a perception out there that everyone can be a safety professional. Then I thought of asking my CEO if he had a robust program to run the company, but I did not have the courage to do so.

We all knew our company was well run based on some core values such as "customer intimacy," etc. The CEO had the people and processes in place to build an outstanding operation. In the same manner, I believe we had outstanding safety processes in place in order to maintain outstanding safety records. But it was still evolving and needed to continually improve just like our business processes and activities. What we needed were processes that are so integrated into operations that safety is seamless. If we were able to make the safety practices at work invisible, we would have succeeded in getting closer to espousing world-class safety.

On the one hand, I liked what the plant safety team leader had said because it made me feel that I was closer to achieving world-class safety as it was invisible, but on the other hand, I felt bad that he had difficulty understanding our safety processes. So I scheduled a special trip to that plant and presented FDC's Corporate Safety Plan to all the managers, including the safety team leader. After my presentation he came to me and said that he now understood the safety process and culture we were trying to achieve. I felt comfortable with his acknowledgement and that we had the proper processes—the Five Pillars of Safety—in place to protect our employees and operation.

Culture and character go hand in hand. They are integral parts of each other. At one of the plants, safety culture was at the top of everyone's minds. Once I had forgotten to bring along my safety shoes when visiting this plant. I was thinking about getting away without the shoes, hoping no one would notice them. I talked to one of my associates in corporate and he said, "Don't even think about it. The employees will check you out as soon as you enter and will ask you to remove your shoes so they can check it out. If they find these are not safety shoes, you will not be allowed to enter even if you are the corporate safety director. And you will be embarrassed and put the plant in the newspaper trying to sneak in."

I asked him what I should do and his advice was to come back or buy new shoes before entering the manufacturing plant. I took his advice and stopped over to purchase a new pair of safety shoes for the plant tour. Sure enough, I was asked to remove my shoes for a thorough examination. That was how strong the safety culture was at this plant. Every employee took safety to heart and looked out for each other.

Companies that practice world-class safety take every opportunity to show that they care about employees like family. The executives find ways to know employee families and help them out on a personal level. Valuing employees and their contributions is their core value. They focus on employees knowing fully well that profit and process improvement will follow suit. They make sure employees share best practices so that everyone improves both at a business level and personal level.

For world-class safety companies, the employees are not only important when they are at work, but also when they are off work! Whatever I learned at work with regards to safety, I practiced it at home because it was passionately being promoted at work. I would make sure my stairs were clear and not cluttered with toys or my safety books for these could create a trip or fall hazard. One of my managers hurt his feet while mowing the lawn. People do stupid things; he somehow pulled the lawnmower backwards into a tight space to cut a small bush instead of pushing the lawnmower forward or using another appropriate garden tool. As a result, he severely hurt his toes. He should have also put on safety shoes while mowing. Each time I go out to mow the lawn now, I always have my safety shoes and safety eye glasses with the side shields as a precaution.

The company placed safety observation posts at the entrance of the road to the plant in order to observe how many employees had safety belts on. This was so the company could encourage the employees to drive safely and think about safety while driving. We provided gifts as a token of our appreciation to employees who followed safety precautions driving their vehicles. For world-class safety companies, it is imperative to promote off-the-job safety. The fact remains that whether a person is injured on the job or off the job, he or she is not available for productive work when that happens.

Developing a good safety culture has a lot to do with the leader of an organization who is responsible for any major decisions being made. Leaders may not fully represent the company culture, but it is often perceived as such. For example, I have heard safety consultants and OSHA inspectors say a company does not care about employee safety. What they really mean is that the current leadership—in most case, it is the top leader—does not care about their safety. I believe this may have a lot to do with a leader's knowledge of safety. If we were able to teach and give the leader an understanding of safety, I believe he would be more likely to proactively promote safety.

In many cases when I made safety recommendations as a consultant, I was told, "we will take care of that." But it never got done. On the flip side, I was involved in a group that organized a safety conference in the local Chicago area. For years they ran the event without liability insurance or even getting incorporated as a non-profit organization. Since I was involved in the purchase of insurance, I knew this was a very critical business function. I would not even think of running a company without adequate insurance protection or a government issued license. In this case, the chairperson running the event did not even know the organization's federal tax ID number or if they had one. The group had an OSHA director and other safety professionals, including Industrial Hygiene (IH) in its various committees. The missing link was the knowledge. They knew how to ensure safety, but not how to run an organization properly. If these safety people had adequate knowledge of business, they would not have functioned without following proper business norms. Again, when I brought this up, the answer was, "We will take care of that." Obviously, it never got taken care of. Developing a safety culture or sound business culture requires knowledge of these subjects.

The Fort Dearborn Company was recognized by *EHS Safety* as one of America's Safest Companies for the year 2005 because, as its editor and associate publisher said, "They understand that work-related injuries and fatalities are a cost in human and financial terms that no company should expect to incur. That's why they apply their management skills, ingenuity and resources to ensuring that their employees are safe on and off the job." It went on to say, "When a printing company stops printing to install a ventilation system, you know safety is a way of doing business."

While installing a new, six-color printing press at one of the company's facilities, it was determined that the ink-mixing room had to be moved to a new area. The new room did not have exhaust ventilation. Ink was necessary to continue production, but proper ventilation was necessary to continue making ink in a safe manner. Production was stopped until ventilation could be installed for employee safety.

"There was no question about continuing production," I had said to management. "We could not put our employees at risk." And they agreed. This instance shows top-quality safety culture at all levels of the work force. Employees respond well to management's commitment to safety if they see management's emphasis on safety is not just talk. Safe and healthy employees are productive employees; that's good business. In fact, the company focus on hazard prevention and injury reduction paid off as the insurance broker said the company saved more than a million dollars since developing continuous improvement programs for positive culture change. That is substantial. Safety does make good business sense. But the most important thing is that you send your people back to their families safe every day. That is what the management believed and that is what they promoted.

The leaders of world-class safety companies understand the need to continuously learn, innovate, and improve. In order to achieve this goal, and due to an aging work force, I invited a professor from Northern Illinois University's graduate program in ergonomics to send his students out to company facilities and analyze ergonomics. The review, which cost the company nothing since it was incorporated into the students' semester studies, focused on material handling issues, since about 70 percent of the company's ergonomic injuries were related to sorting and palletizing.

The suggestions from the ergonomics students included buying vacuum lifts and lift assists to reduce bending. The company purchased new equipment and, at the same time, trained managers to identify ergonomic hazards before they become an issue and employees get hurt. For example, a customer wanted the company to pack labels in a box and the box weighed 60 pounds. Company guidelines stated that employees could not lift anything above 35 pounds. The manager stepped in and told the customer, "We can't do this because we went through ergo training and we know it's not safe for our associates to lift 60-pound boxes."

The customer agreed to change its requirements from 60-pound boxes to 35-pound boxes. The result was the boxes were easier to lift, thereby reducing the potential for injury not only for company employees, but for anyone who has to move those boxes further down the line in the customer facilities. Again, world-class safety companies try to make safety contagious.

Not everyone believes that organizational culture influences WCSC, however. An article by Dr. David Michaels called "Changing Safety Culture at America's Leading Furniture Manufacture" depicts this belief. According to him, a company can demonstrate its commitment to safety for its employees through an "agreement with OSHA" and by complying with any citations issued by its compliance officers, which require the correction of all violations or the payment of a significant penalty of $1.75 million. The written agreement with the company and threat of significant penalties is a catalyst for making a meaningful change.

To this end, OSHA promotes its Voluntary Protection Program (VPP). When a company approves this program, in the eyes of OSHA, it recognizes outstanding, exemplary world-class safety culture. Each company site must apply for VPP status followed by rigorous onsite evaluation by a team of OSHA safety and health experts. The sites are required to have written safety and health programs in place with ongoing documentation showing that the company is working with OSHA in implementing government regulations. Once this is verified by a follow-up inspection, OSHA provides the site with blessings and a flag showing the company is the best in the crowd.

Most safety professionals believe that company culture has an impact on safety. We should want to do safety initiatives because it is the right thing to do. We do it right all the time with a high degree of reliability.

Another factor that affects company safety culture is the legal and litigation environment we have in the U.S. All employees from senior management to floor helpers need to understand the relationship between safety and the law. One of our plants had a strong caring attitude for its employees. So when there was an injury, a thorough investigation would identify what people in the company could have done to prevent the incident. One of our customers making an onsite delivery claimed a couple of days later that he fell on the ice in the parking lot. The manager of the plant tried to complete a thorough investigation using techniques he would apply for employee injuries not realizing the legal liability implications. A liability investigation by the corporate team found the snow had fallen the day after he claimed his injury on our parking lot! So company safety culture should also educate key employees so that they understand the legality of various situations.

With dedication and patience, the safety professionals in

an organization can help the organization achieve world-class safety culture by encouraging the senior-level management to work together with employees and OSHA in its safety initiatives. Management, employees, and safety professionals need to understand the local and federal laws that apply to make this a seamless effort.

Management Leadership and Commitment

As a safety expert who has worked in the industry for more than forty-four years, I believe management's commitment to safety is the single most important factor in achieving safety results. Senior management is usually an individual or a team who are held responsible for the effective performance of the operations of the company. Senior management's commitment to safety in a privately-held, small to medium-size corporation can be obtained by pointing out the financial benefits of the program, since the prime driver for these companies is the ability to make a profit. Although profit is a motivator for public corporations, other factors, such as social responsibility, play a significant role in senior management's decision-making process.

What does the word commitment mean? To commit to something typically demonstrates a person or organization's strong sense of intention and focus. It can be accompanied by a statement of purpose such as a mission statement or a plan of action. Very often it is meant to show the seriousness of one's relationship to and involvement in safety. This commitment should come from the bottom of one's heart.

When I was a loss-control consultant, I pointed out this deficiency to an agency account executive and, in order to retain business and keep everyone happy, he put together a mission statement or a plan of action with the top company executive name and sent it to the underwriter. Was this senior executive really committed? When I met with this executive and asked him about this commitment, he could not even remember the statement he had made. On the other hand, I worked for an executive who showed up at the plant on a day when it had snowed eight inches or more. All non-essential plant employees were asked to take the day off. He was there just to make sure the emergency crew was working safely on what they needed to do in order to protect the plant operation. Now that is commitment! That is what a world-class company senior executive would do.

Visible senior management leadership and commitment to safety is one of the most important factors in developing the culture necessary for achieving the goal of being a world-class organization. Here I will discuss how senior management's commitment to safety can be secured and how to use financial and other matrices to measure the level of their commitment. The techniques discussed include:

- Appealing to senior management's values
- Generating financial benefits from safety initiatives
- Demonstrating the cost of non-compliance to safety standards
- Developing and instituting an effective safety system

A proactive safety professional should understand the perspective of their company's senior management and what distinguishes the organization from others in the same arena. Some senior management leaders take a holistic approach and may require constant communication. Then there are those who make knowledge-based decisions; they need to understand the logical reasoning behind your safety-related activities and expenditures. All of them want their organizations to succeed financially and to be the best they can be. So as safety professionals, we need to make sure our safety initiatives reap a financial benefit for the organization. Successful organizations have great financial, production, and quality-assurance systems in place and they expect to have excellent safety systems operating as well.[1]

Although many companies may be having regular safety meetings and completing OSHA required safety training, they still find the cost of injuries going up.[2] The missing link is a firm commitment, preferably a highly visible one, to safety from senior executives and managers.

A company's senior management may want to change their track record of injuries and create a positive environment for a cultural change. However, they are often not sure how this can be accomplished. The motivating factors behind the desire to positively impact their safety culture are the requirements of the law and the financial benefits to the company. Since these senior managers have a significant influence on the organization's safety culture, they need to continuously demonstrate a visible commitment to safety, best indicated by "the proportion of resources (time, money, people) and support allocated to health and safety management and by the status given to health and safety."[3] How do you go about getting senior management to visibly commit to safety now that you know what will motivate them?

---

1   Michael Weibert and Catherine Plunkett, "Motivational Safety System: One Organization's Experience Moving Toward World-class Performance," *Professional Safety*, February 2006: 34–39.
2   U.S. Department of Labor, "Occupational Safety and Health Act of 1970," *General Industry Standards*, vol. 6, chap. XVII, 29 CFR pt. 1910, Washington, D.C.
3   R. Flin and S. Yule, "Leadership for Safety: Industrial Experience," *Quality Saf Health Care*, 2004. http://qualitysafety.bmj.com/content/13/suppl_2/ii45.full

The High Cost of Non-compliance to OSHA Safety Standards

As safety professionals, it is our task to help managers realize there is a direct and positive correlation between investment in safety and the subsequent return on that investment. There is evidence that companies that implement effective safety programs can reduce injury and illness rates by more than 20 percent and generate a return of more than four dollars for every one dollar invested in safety. This also allows companies to avoid OSHA penalties in the case of an audit. Safety is a non-delegable duty; per OSHA directive, senior executives and general managers cannot delegate safety responsibilities to line supervisors. The expectations of various governing bodies on private organizations in the U.S. (and several other developed countries) have increased significantly. Safety regulations have become more complex with the rise of criminal investigations for non-compliance and the advent of significant OSHA fines. Let us take a look at OSHA's penalty structure:[4]

Citations and Penalties

Violation Type	Penalty
WILLFUL A violation that the employer intentionally and knowingly commits or a violation that the employer commits with plain indifference to the law.	OSHA may propose penalties of up to $70,000 for each willful violation, with a minimum penalty of $5,000 for each willful violation.
SERIOUS A violation where there is substantial probability that death or serious physical harm could result and that the employer knew, or should have known, of the hazard.	There is mandatory penalty for serious violations which may be up to $7,000.

4　U.S. Department of Labor, Occupational Safety and Health Administration, "OSHA Inspections," OSHA 2098, 2002 (revised), accessed April 13, 2015. https://www.osha.gov/Publications/osha2098.pdf

Violation Type	Penalty
OTHER-THAN SERIOUS A violation that has a direct relationship to safety and health, but probably would not cause death or serious physical harm.	OSHA may propose a penalty of up to $7,000 for each other-than-serious violation.
REPEATED A violation that is the same or similar to a previous violation.	OSHA may propose penalties of up to $70,000 for each repeated violation.

With professional safety guidance from in-house or private safety consultants, a company's bottom line may be significantly strengthened by compliance to safety and OSHA standards.

The cost of a willful violation—a violation that the employer allegedly intentionally and knowingly committed—may cost up to $70,000 for each violation. A company and the management individuals responsible for safety, which includes general managers and department managers, can be fined up to $500,000 and six months in jail if convicted in criminal proceedings of a willful violation as shown in *Secretary of Labor v. E Smalis Painting Co., Inc.*[5] General managers and supervisors, as "agents" of the employer who have the authority to create legal liability for the company, can be personally held liable for monetary judgment against them and their financial assets or worse, face personal criminal liability with jail terms. This is a strong message from the Occupational Safety and Health Administration (OSHA) encouraging companies to develop a proactive safety culture.

Please use the guidelines below to prepare for and manage an OSHA inspection if and when you are audited.

An OSHA inspection can be triggered by any one of the following:

---

5   Secretary of Labor v. E Smalis Painting Co., Inc., Occupational Safety and Health Review Comission, Docket No. 94-1979, April 10, 2009. http://www.oshrc.gov/decisions/pdf_2009/94-1979.pdf

1. Imminent danger: allegations of an imminent danger situation will receive highest priority. The inspection will be conducted within twenty-four hours of the allegation unless extraordinary circumstances exist.

2. Fatality and catastrophe: accidents will be investigated by OSHA within twenty-four hours if they include any of the following conditions:
   a. One or more fatalities
   b. Three or more employees hospitalized for more than twenty-four hours

3. Investigation resulting from a national office special program.

4. Employee or ex-employee complaint: complaints are investigated by inspection or by letter (ex-employee complaint only) and a potential follow-up inspection.

5. Programmed inspection: policy requires that programmed inspections are conducted in industries where injury rates are high.

6. Follow-up inspection: OSHA can re-inspect to assure that an employer has abated the violations that have been cited. Fines are usually approximately ten times higher for "Failure to Abate" citations.

If an inspector does not have a warrant, the Fourth Amendment of the Constitution allows you to deny the inspector access to the facility. It is your decision whether to require a warrant or voluntarily consent to an inspection; I would recommend voluntarily consenting to the inspection.

OSHA Audit Preparation: to prepare for an OSHA inspection, you should designate a representative, generally the general manager (safety team leader, HR manager) prior to the inspector's arrival. The inspector's credentials will bear a photograph and serial number with the nearest OSHA office and should be checked; he or she should be accompanied at all times by the designated representative. If at any time the representative has difficulty responding to a question, he or she should telephone for advice from an attorney or a trusted, knowledgeable source.

Once the inspector is in, the protocol that he will follow will consist of:

1. Opening conference
2. A tour of the facility
3. Closing conference

Opening Conference: during the opening conference, which usually takes about an hour, a number of topics will be discussed, including general questions about the business. He or she will review the purpose for the visit (e.g. if there is a complaint, accident or program) along with the scope of the inspection (part of the facility or wall-to-wall). A copy of the complaint will be given, if one is involved. Please make a note of why you were selected and what is going to be inspected.

Handouts of OSHA pamphlets will be given to you. Please keep all of them with the date and name of the inspector who issued them.

You will need to identify trade secrets if there are any. Please note that any records that are not specified on the warrant or in his or her request do not have to be provided.

The inspector will generally review the following records or programs:

• The Hazard Communication written program for your facility. These include provisions for labeling, material safety data sheets, employee training, and a list of hazardous chemicals.
• The Lock Out Tag Out written program will be reviewed and you may be questioned as to your knowledge of the program.
• The Injury and Illness Log (OSHA Form 300) will be inspected.
• Safety programs will be checked to see if they are being observed.

Tour of the Facility: the route and duration of the tour is determined by the inspector and accompanying representatives. If the inspector wants to see a specific spot, take the inspector there directly, rather than walking through the entire plant. A detailed observation of the facility by the inspector can include talking with employees, note taking, making instrument readings, taking photos or using a video camera.

Never leave the inspector alone, and do not volunteer any information. Make sure department managers know to answer all questions honestly without volunteering extra information. The inspector may consult with employees as long as it does not interfere with work operations and the employee does not object. He or she may also meet with the employee in private if the employee does not object.

During the tour the inspector may point out things he "believes" are in violation of the OSHA Act. If you agree they are violations, you will surely be cited and fined. If you are able to correct conditions on the spot, do so, but you may still receive a citation and penalty.

If the officer takes notes or measurements, uses a camera or video tapes, you should do likewise. Record everything that happens, including the time and date.

What you say can and will be used against you. Never engage in idle talk or chit-chat.

Never give estimates if you do not have accurate information. You may be providing OSHA with false information, which is a criminal offense.

At the end of each day's inspection, go over your notes and measurements for accuracy and completeness.

Closing Conference: at the conclusion of the tour, the inspector will hold a conference. The purpose of this is:

- to advise you of the conditions observed in the facility
- to obtain further information

- to relate any possible citations that may be issued, your right to appeal, and time limits
- to answer your questions

If any violations were voluntarily corrected on the spot, it is essential that the inspector states that the violation was abated before he leaves the premises and notes it with the date, time, place, and a witness present.

You may be asked how long it will take and how much it will cost to correct a citation. Do not agree that they are violations, for you may be held liable by how you respond. If it is an obvious violation, providing information may help OSHA determine the time needed for abatement.

The inspector does not propose penalties. The U.S. Department of Labor Area Director will notify you in writing by certified mail of any citations or penalties received. You have fifteen working days to either pay the penalties or contest the citation, the penalties, or both. Failure to contest the citation confirms the penalty as final.

Violations, Classifications, and Penalties: there are different classifications of violations and penalties structured for the workplace: "Other than Serious Citation," "Serious Violation," "Willful Violation," and "Repeat Violation." This has been discussed in the table earlier.

In summary, to survive an OSHA inspection, ALWAYS BE PREPARED. Understand the law (consult 29 CFR 1910, General Industry Standards), have a self-inspection program in place, and keep the following written programs available for inspection: General Safety Program, Hazardous Communication Program, Lock Out Tag Out Program, OSHA required forms such as OSHA 300 forms and other pertinent safety and health related programs in place.

Improvement in safety improves productivity and mitigates fines because employees who feel safe often work harder. For example, when working for an insurance company as a loss-control consultant, I remember visiting a glove manufacturer that had one finger amputation every year for several years when employees were cutting cloth with foot operated clicker presses. From the perspective of the general manager at the time, it was simply the cost of doing business until the insurance company recommended that the company implement two-hand trip devices. Once the devices were implemented, production went up by more than 20 percent as the employees no longer had to worry about accidentally amputating a finger. Improvement in safety also improved financial performance and minimized operational risks for the company. The safety of employees while on the company's premises is not only the company's moral and legal obligation, but it also makes good business sense.

Allocating the Cost of Injuries for Cultural Change

Requiring management's visible commitment to safety is not an easy task. Another method that can be effectively used to obtain the commitment of general managers is to proactively promote safety and control the cost of workers' compensation by allocating the cost of injuries back to each of the plants or profit centers. Different systems can be used to allocate costs in a way that is fair and equitable. Most companies charge back the cost of workers' compensation to the plant based on its payroll, a system which is not sensitive to controlling injuries and associated costs. Some companies pay the workers' compensation premiums up front, but collect monies from each plant after the end of the policy year based on the percent cost of injuries per plant per year. This change in allocating the cost of injuries is an improvement since it encourages a proactive safety process in order to control employee injuries and their costs. A better system is to collect the cost of workers' compensation insurance from each plant using a formula and then provide a rebate after the close of the policy year based on plant injury experience. The formula should have two components: a fixed cost and a variable cost. The fixed cost for each plant should be the company's deductible premium for the large loss deductible program divided by the total company payroll multiplied by the plant's payroll. The variable cost should be the weighted cost of injuries measured by the incurred cost for the last four years. The variable cost of injuries could then be allocated based on a four-year-average-allocation percent developed per plant. An example of variable cost allocated might look like this:

Allocation of Workers' Compensation Insurance Costs

Policy Year: Current **PLANT ALLOCATIONS** *Claim Year Losses Updated Using April Loss Runs*							
	Total Incurred Policy Year				Weighted Average Loss	Loss Allocation Percent	Allocation of Projected Losses
	A	B	C	D			
Annual Weight	10%	10%	40%	40%			
Plant 1	$395,395	$242,288	$114,687	$35,305	$123,765	32.23%	$253,313
Plant 2	$135,835	$320,610	$25,162	$63,741	$81,206	21.15%	$166,206
Plant 3	$19,280	$2,668	0	$72,691	$31,271	8.14%	$64,004
Plant 4	$27,115	$100,933	$215,118	$19,963	$106,837	27.82%	$218,666
Plant 5	$543	$282,441	$1,861	0	$29,043	7.56%	$59,443
Plant 6	$96,319	$3,529	$1,103	$3,578	$11,857	3.09%	$24,268
Corporate Office					0	0.00%	
Gross Allocated Cost	$674,487	$952,469	$357,931	$195,278	$383,979	100.00%	$785,900

This system of allocating the cost of employee injuries through workers' compensation is fair and equitable and requires the active participation of all general managers and employees to secure safety in the plant operation. It is often critical for the successful operation of safety at each plant and, in many cases, determines the overall success and bonuses earned by each of the plant general managers.

There are other allocation methods available for allocating injury cost in a fair and equitable manner. Instead of using the incurred cost of injuries to determine the weighted-average-allocation per-centage, one could use the OSHA injury rates as follows:

First of all, determine the Incidence Rate (IR), Days Away Restricted Transferred (DART), and Lost Work Day Incidence (LWDI) rates you would like to use for calculating the weighted average for each of the plants.

Allocation of Workers' Compensation Insurance Costs

Calendar Year							
	# of Employees	# of Incidents	Incidents with Restricted or Transfer	Lost Time Incidents	IR	DART	LWDI
Plant 1	222	12	2	1	5.41	0.90	0.45
Plant 2	156	9	2	2	5.77	1.28	1.28
Plant 3	119	3	0	0	2.52	0.00	0.00
Plant 4	72	8	3	0	11.11	4.17	0.00
Plant 5	29	3	1	2	10.34	3.45	6.90
Plant 6	49	3	2	1	6.12	4.08	2.04
Corporate	80	0	0	0	0.00	0.00	0.00
Total	727	38	10	6	5.23	1.38	0.83

Calendar Year						
	# of Employees	IR	DART	LWDI	Weighted Average	Loss Allocation Percent
Annual Weight		20%	40%	40%		
Plant 1	222	5.41	0.90	0.45	1.62	9%
Plant 2	156	5.77	1.28	1.28	2.18	12%
Plant 3	119	2.52	0.00	0.00	0.50	3%
Plant 4	72	11.11	4.17	0.00	3.89	22%
Plant 5	29	10.34	3.45	6.90	6.21	22%
Plant 6	49	6.12	4.08	2.04	3.67	20%
Corporate	80	0.00	0.00	0.00	0.00	0%
Total	727				18.07	100%

The variable workers' compensation cost for the following year would then be distributed based on the percentages calculated as above. You will notice the small plant with only twenty-nine employees had to bear 22 percent of the entire cost using this method because it had a lot more injuries this year than expected. In order to minimize the significant impact of a poor injury experience from one year to the next, you can use a weighted average for the last three or four years.

Instead of using the weighted average for IR, DART, and LWDI, you can use the DART or Lost Workday Incidence and Illness (LWDII) rates only to allocate the cost of injuries for the following year. You can also use the Experience Modification (MOD) rate for each plant if it is available. Generally, for a company with several plants, the experience MOD is the composite MOD for all the plants. In such cases there are programs that will help you develop an experience modification for each plant, which can then be used to allocate the cost of injuries for each plant.

However, in my experience, using the incurred cost method discussed earlier tends to work out much better for every plant.

Return on Investment (ROI) – Cost-benefit Analysis[6]

Senior management's commitment can be secured for new safety projects if they receive calculations showing the Return on Investment (ROI) and how the company will benefit from investing in a specific safety project. ROI is the financial measurement commonly used to evaluate the attractiveness of one investment over another. Most companies require ROI calculations for all investments, and such calculations should pass a hurdle rate or minimum rate of return (often

6   David S. Pais, "Showing EHS Value Through Return on Investment (ROI)," presentation at the ASSE Professional Development Conference and Exposition Safety 2011, June 15, 2011, Chicago.

8 to 9 percent) in order to be acceptable. According to David Pais, "Many companies establish a hurdle rate of 12 percent, the minimum rate of return on a project to determine if the investment passes the financial test."

For example, take a printing company that was having significant ergonomic injuries in its finishing department. The finishing department had five paper-cutting machines that required employees to cut paper using two-hand trip devices. Once the cutting cycle was complete, employees would pick up the waste paper trims with their right hand and throw the trimmed paper in a cardboard box placed behind them.

Later, when they had time, they or a helper would pick up the waste trim from the floor and place it inside the cardboard box. Once the box was full, a lift-truck operator would dump it in a baler. The bales of paper trims would then be picked up by a waste hauler for recycling.

This process required awkward posture and shoulder movements by the paper cutters, often causing severe back and shoulder injuries. It also required the helpers to bend down and pick up the waste trim, which caused some back injuries. In addition, the lift-truck operator could potentially hurt the finishing operators while picking up the boxes of waste trim.

The safety manager analyzed the process and recommended an automatic vacuum trim system that would pick up the trims from all five finishing cutters and move them to the baler via air ducts. However, the cost of the system was $250,000, and it was critical to show the financial benefit by completing the ROI calculation:

ROI = Benefits/Time Divided by Initial Investment

For most Safety Projects,

ROI = Risk Reduction Divided by Cost

However, for this specific project there were significant benefits and risk reduction associated with the safety system, so ROI was calculated as:

ROI = (Benefits/Time) + (Risk Reduction/Time) / The Initial Investment

The table below shows the benefits over time and savings in risk reduction.

Savings Realized Per Year

Year	Item	Benefits	Cost Reduction	Total Savings
1	Helper not needed ($12.28 per hour plus benefits at 23% of rate – Two shifts)		$60,418	$60,418
2	Increase cutter efficiency – 4%	$12,084		$12,084
3	Reduced Waste	$1,200		$1,200
4	Savings in Leasing Lift Truck		$2,178	$2,178
5	Reduced Cost of Ergonomic Injuries		$25,360	$25,360
6	Good Night Sleep – Increase level of safety confidence	$5,000		$5,000
	Total			$106,240

A "good night's sleep" is the risk management concept that refers to the value senior management assigns to employees, their families, and management sleeping well as a result of a safety-related improvement. This value is higher if the potential for loss is high without the intervention of a safety program. This is a subjective number and varies depending upon the exposure, reduction in potential for loss, and the senior management affected.

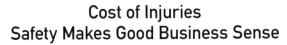

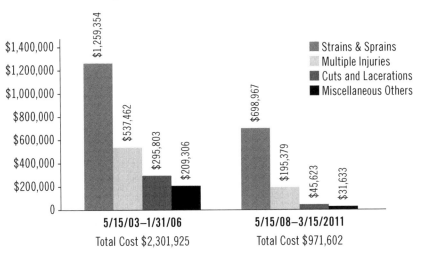

For a three-year period, based on insurance company loss analysis, the company saved $560,387 throughout the six plants, with a projected savings of approximately $31,132 for the specific plant in question and an estimated savings of $25,360 in the finishing department.

ROI for this project was calculated as $106,240/$250,000 = 42 percent every year. In other words, it would take about two years and four months to recoup the investment. The three-year rate of return was calculated at 27 percent [($106,240 × 3 − $250,000)/$250,000], handily surpassing the annual company hurdle rate of 8.5 percent. If the annual hurdle rate (minimum acceptable rate of return) was higher than 9 percent, then this project would have been denied if the decision was made purely on a financial basis.

There are various factors that can change the ROI calculations:

- The hurdle rate for companies whose cost of capital is high tends to be higher.
- The years needed to recoup the investment may vary depending upon the type of investment and the company's tax amortization policy.
- The Present Value Interest Factors (PVIF) for multiple year rates of return if you would like the net present value included in the calculations. This calculation is based on the concept that the amount of money received today is worth more than if the money is received in the future and is generally completed using an appropriate software program.

Each of these points (high cost of non-compliance to OSHA, allocation of cost of injuries, ROI, etc.) can be used to effectively gain senior management's commitment and leadership since they have a higher level of impact on their performance. When management is financially proactive in encouraging safety activities, they demonstrate a commitment to cultivating safety culture that is found only in world-class safety companies.

When I was a loss-control consultant I reviewed an auto dealership. We had made several safety-related recommendations during our initial visit, so I visited the site again to determine whether they had complied with them or not. The CFO, my contact, had sent a letter to me stating they had complied with all of the recommendations. He also said as much to me when I met him in the office during the re-inspection, and he suggested that I talk to the service director. When I met with the service director, he said he had given the recommendations letter to the maintenance porter for completion. We could not find the maintenance porter, but his coworker said he put the letter along with every record on maintenance relating to safety in the "safety box container." When we looked in the safety box—a file box with all items related to safety—we did not find the letter showing that the maintenance recommended had been complied with in.

Nobody in the dealership had taken responsibility for completing the safety-related recommendations. I did not know how to convey the message that someone in management needed to take on that responsibility. So I took out the handkerchief I used to clean my runny nose and dropped it on their clean floor. Guess what happened? Everyone in the company, including the president, came to know that a safety inspector had dropped a dirty handkerchief on the floor. Commands went out to the general manager and to the service manager that this should be cleaned up right away. A porter was sent in less than ten minutes to clean up the floor and keep it spotless. The president does not like dirty floors because the dealership corporate does not like dirty floors! What would happen if the president cared as much about safety and did not like safety-related issues to be left open?

Let me share another similar experience. While working at an insurance brokerage firm, I made an insurance and loss-control presentation some fourteen years ago at FDC's executive board meeting on safety, as its loss experience was very poor. After discussing what the company could do with regard to improving safety programs, something struck my brain. I asked the CFO what his background was and he said he had MS in accounting with significant experience in accounting and finance. I asked the manufacturing VP his experience and he said he had an MBA and that he had been running the company successfully for many years. I asked the quality control manager; he said he had an MS in physics and had studied quality control extensively. Then I asked the safety person who was invited to this meeting what his experience was and he said he was in maintenance and volunteered to manage safety in his free time.

I asked the CEO if he would hire volunteers to run the plant or complete his books or run the QC Department? He said "no." I raised the question that if a volunteer runs the safety department and he has no safety education, how can the CEO expect safety records to improve? After this meeting, HR called me and asked me to be on board. Perhaps they were thinking: "Big mouth, let us see what you can do?" Cutting a long story short, I accepted the challenge of overhauling their safety program and got to work. Under my supervision, the company was named one of the safest companies in the U.S in 2005 by Environmental Health and Safety. The rest is history. When you hire a professional dedicated to safety, you will get results—it just happens. You put in place a program and procedure that will bring you close to world-class safety and in the process you protect business interests, employees, and keep the OSHA at bay.

Both these examples show how one can make companies and senior employees take crucial accountability for safety and developing a world-class safety culture. Various other methods on how to achieve this are discussed in the book *Crucial Accountability* by Patters, Grenny, Maxfield, McMillan and Switzler.

Not only is it important to get every employee, including senior executives, actively engaged in company safety initiatives, but it is also just as important to make sure you yourself are actively engaged on a daily basis. I am a somewhat religious person, and my parents taught me to pray at least once a day. So every day in the morning I used to pray for the safety of my employees and the company as a whole. I am not sure if my prayers worked; though, I believe they did. What I know for sure is that the daily prayers kept me thinking about my employees' safety every day and encouraged me to work on developing new safety initiatives for their safety.

When I came to America, my mom told me, "If you are in any difficulty, just close your eyes and pray and things will work out just fine." Back home I was used to driving on the left side of the road. When I got my first job with American Mutual Insurance Company, I got a new car and had to start driving on the right side of the road. Generally, it is fine to make the switch as long as you are following the vehicle in front of you. However, once I was alone, I reverted to driving on the other side of the road. Not long after that I saw a car coming my way. I did not know what to do. It is then that I remembered my mom's golden rule: Close your eyes and pray. (P.S. I do not recommend this rule when driving.) I just did that and when I opened my eyes, the road was clear. A couple of years after that incident, I was discussing Defensive Driver Training to a group of drivers and told them my golden rule. A big guy at the back of the room got up and said, "I have been looking for you for all these years! I was in the ditch avoiding you!"

In a world-class company, safety should always be included in the company planning process and for measuring success. The company I worked for used a program called Snapshot. Every leader had to prepare a monthly snapshot of where they were and where they were heading using the following categories: Goals and Objectives, Current Snapshot, Performance, Green/Red Flags and Action items. Below is an example of a safety snapshot. It was a very effective program in getting all managers actively engaged in safety activities and keeping them informed about progress and problems.

Safety Snapshot

Leader: Mike Saujani	Date: 3/13/03
Team Members:	

Goals & Objectives – *Keep the big picture in mind*

1. Zero safety violations
2. Full management and association participation
3. Fully functional safety committees
4. Zero Lost Time accidents

Current Snapshot – *Summarize the project status*

We had a Safety Team Leaders meeting in March. We discussed losses, successes, property conservation programs, accident investigation; lock out tag out programs and return to work programs. The feedback I got was very positive. The only negative comment was that we tried to do too many things in a day.

The managers are completing daily inspections and I believe we have reached or closed to reaching the goal of no violations. The Illinois State inspectors plan to visit our facilities within 30 days to confirm this.

All divisions are now having monthly Safety Committee meetings and developing action plans to improve safety and associate participation.

Various insurance companies should visit our facilities for an evaluation and bid on our insurance programs. My perception is that the agencies are pleased with what we have been able to accomplish this last year. We need to see it being reflected in our insurance costs. These costs have gone down because of our self-insured program and the General Managers being held accountable for safety.

We are working on zero lost time accident by improving ergonomics and encouraging return-to-work program.

Generally, I feel, we should be proud of our accomplishments.

Performance – *What do the numbers tell you?*

Measure	Target	Current Week	MTD Total	Week of
Incidence rate of 2.7 is well below the industry average of 5.1				
LWDII rate of 1.1 is well below the industry average of 2.3	0			
Incurred losses of $480,751 for eight months is at target	$720,000			

Green/Red Flags – *What are the corresponding action items?*

G1. Identified Hazards are corrected immediately.
G2. Very good participation by associates and managers in safety.
G3. Safety Committee meetings are being held and minutes completed.
G4. Safety programs are consistently being followed at all divisions.
R1. Develop and implement Return To Work programs to achieve zero lost time accident goal.
R2.

Action Items – *R1*	Assigned to	Due Date
1. Develop return-to-work program	Mike/Bill	3/4/03
2. Train safety team and HR leaders	Mike/Bill	3/4/03

3. Train managers	Mike	6/1/03
4. Implement the program	HR	Continuous
5. Monitor the program	Mike/Bill	Continuous
6. [Add type of action item here]		

Do you need to redistribute resources to meet your deadlines?
Minimize the number of open action items to ensure progress.

Purpose

Snapshots provide a systematic way of measuring and communicating progress and performance on any given program or project.

Red/Green Flags

Red and green flags should reflect high-level achievements and issues, summarizing the events of the week and the most critical areas of concern. It is imperative that each red flag have an associated action item assigned to it. In the case of green flags, please list any parties that should be notified of the accomplishment.

I had the uncanny ability to foresee what could happen if safety precautions were not complied with. I believe that is what my various employers saw when hiring me, otherwise I would have been let go several times during my career. The first few weeks at each of the company I worked for were typically difficult for me. Maybe because my stress level was always high when starting a new responsibility. The first time I was hired as a safety engineer at American Mutual Insurance Company, I was provided with a new vehicle—a Grand Torino. The second day on the job, I was stopped behind a vehicle making a right turn. When the light turned, the vehicle in front started moving then suddenly stopped. I rear-ended the vehicle. At Hanover Insurance Company, while my superior was introducing me to the management team, I accidentally hit a glass of water and spilled it on him! Two weeks after I was hired by Amerisure Insurance Company. I had to drive to its corporate office in Farmington Hills, Michigan for training. While backing out at a gas station on my way to Farmington Hills, I hit the gas pump protective barrier and damaged the company car. When I was hired as safety director at FDC, the CEO took me out for lunch on my first day at work. I drove us to lunch, and on the way back to the plant I parked in the spot next to his brand new BMW. While getting out, I put a nick in his car! When I look back, I feel I probably should have been let go after each of these incidents. Lucky for the company and me, I was not. I had the ability to foresee what needed to be done to protect the company employees and its operations.

While surveying a powder coating operation in Chicago for Hanover who insured the building for an investor, I noticed several areas where the electrical wiring was not in compliance with the National Electrical Code. I recommended that these be altered. The company executive told me that the electrical wiring had been there for years, and he wrote to the insurance company stating that he would be responsible if there was any accident caused as a result of the electrical wiring. Four months later there was a fire that destroyed that section of the building and the cause was faulty electrical wiring!

Hartford Insurance Company insured a large construction company building a couple million gallons of underground reservoir for storing water that would get pumped from lake Michigan to the North Western suburbs. It was completed and had to be tested for water leaks before backfilling the excavated dirt. If the reservoir was filled with water for testing purposes, there was a possibility of it being lifted due to buoyancy. When brought to the attention of the company, it ignored that warning and insisted, "We know what we are doing." Sure enough, the structure buoyed when tested and the employees working on top of the structure had to scramble for safety. The construction company incurred several million dollars in losses as a result of not listening to loss-prevention professionals.

When I was at FDC, we were in the process of buying large printing presses for several million dollars. The deal was to be closed in four days. When I came to know of this acquisition, I met with the CEO and explained that if there was an accident on the press while in seller's care and custody, we would be held responsible without an appropriate agreement. His initial reaction was, "Mike, we only have a few days and I do not want to jeopardize the acquisition." He worked diligently and secured an agreement with a COI that insured FDC if anything were to happen on the presses or to the presses while in the seller's care. A month later, before the presses could be moved to the new FDC location, there was a serious employee injury on one of the presses. However, since the CEO listened, a large lawsuit was avoided and saved the company a significant amount of money.

The lesson to be learned is that if the senior executives listen to safety professionals, they can protect their employees and business operations from significant losses.

3

Employee Involvement and Ownership

The companies that are world class or striving to be world class have their entire organizations involved in safety. However, the levels of safety activities that employees and management get involved in are different. In some world-class safety organizations, senior management is actively engaged, while in others employees are taking an active interest in safety. In order for everyone at a company to be actively engaged in safety in some way, it is important to develop and implement several joint safety-related initiatives. The initiatives I have seen successfully implemented by world-class companies include:

- Roles and responsibilities for various roles in the organization (Saujani and Adler)
- Safety awareness surveys to gauge the knowledge of employees
- Encouraging and rewarding safe behaviors
- Engaging associates through:
  - Safety committee meetings[7]

---

7 Katherine Torres, "Making a Safety Committee Work for You," *Occupational Hazards*, October 23, 2006.

- Instituting a safety board program
- Identifying a hazard program
- Designating a Safest Plant of the Year Award
- Planting the seed that they are one of the best companies when it comes to safety; this statement then becomes a reality

Safety is everyone's responsibility. To be successful, it requires the full participation of shareholders, executives, managers, and union and nonunion employees.[8] It is important to have defined roles and responsibilities for each of the stakeholders so that they are all involved in a singular path toward achieving safety success. For example, the role and responsibility of a safety director could be to plan and implement company safety policy, coordinate companywide safety initiatives, audit company facilities and operation, assist in the investigation of accidents, support and encourage safety team leaders, train and motivate associates, and champion safety by engaging associates at the executive level.

The roles and responsibilities of the CEO were to furnish each associate with a place of work free from all recognized hazards. Our CEO went a step further. He made it a point to discuss our safety initiatives and performance at the board of directors meetings. He talked about how safety initiatives positively affected the company's overall performance. His enthusiasm for safety became contagious. After that, the executive board started requiring that all company CEOs report safety initiatives during their presentations. In turn, these CEOs then started encouraging safety initiatives at companies they were responsible for.

Safety became contagious because of the initiative of our CEO. Not only did he promote safety at the board of directors meetings, but he also made sure to follow through the safety requirements at all of the facilities he was responsible for. He made it a point to have

---

8   David A. Jones, "Mapping Support for an EHS Management System," *Occupational Hazards,* June 29, 2006.

his safety shoes and safety eye glasses on when visiting company facilities. Employees noticed and recognized that safety was an important driver of the business.

Associates at all levels can be actively engaged through attending safety committee meetings. These meetings should be fully functional and held at least once a month with a meaningful agenda and list of accomplishments. Employees at this level can be further encouraged to get involved by volunteering for semi-monthly plant safety audits that help identify and assist in correcting hazards and unsafe behaviors. The use of a written checklist to complete the audit is important. It will ensure that areas that need to be inspected do not get forgotten and observations that need to be checked out are not pushed to the wayside.

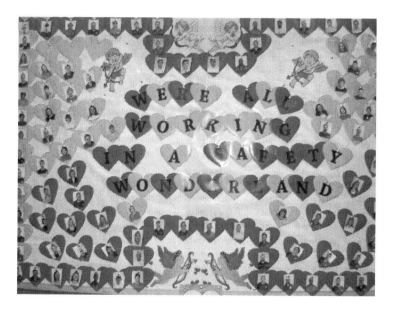

I observed one printing company effectively use the safety board program to garner employee ownership and involvement in safety. The company required employees in each of its four departments to prepare the safety board every quarter with a particular safety theme, such as electrical safety, hazard communication, machine guarding, lift-truck operator safety, etc. The employees worked as a team using their individual expertise to prepare a powerful and innovative board. A team of three senior executives judged the board and the employees in the department with the winning safety board would win a free lunch. The program led to extensive chatter and discussions on safety among employees, which ultimately lead to an outstanding safety record.

An offshoot of the safety board program is the Safest Plant of the Year Award program, which recognizes and encourages plant management to take an active interest in safety. Each plant is required to submit a package of documentation showing their initiatives and successes. The Safest Plant of the Year selection committee consisted of the senior VP of operations, corporate safety director, and the corporate HR director who reviewed the documentation provided by each of the plants. The committee selects the Safest Plant of the Year Award based on the following criteria:

- Results and accomplishments – 30 points
- General managers' commitment to safety – 15 points
- Department managers' involvement in safety – 15 Points

- Fully functional safety committees – 10 Points
- Training and motivation of associates and managers – 25 points
- Miscellaneous – 5 points

Programs like these plant the seed in the minds of plant management and employees that they are the best in the corporation. Once you extol them for being the best, they magically become the best. It just happens.

Monthly Phone Conferences

In order to get safety champions and HR leaders engaged on a regular basis at FDC, we started holding monthly safety conferences. These conferences were held on the third Monday of a month at a fixed time in the afternoon. We always discussed any incident at one of the plants and the action that should have been taken to prevent it. These conferences allowed plants to discuss and share best practices so each of the plants could grow together and learn from the same incidents. For example, at one plant an employee got his finger caught in between rollers while trying to clean them. The interlock limit switch in front of the in-running nip point apparently failed to operate when accidentally touched. After that was discovered, the plant started testing all the limit and safety switches on each of the presses. This, they believed, would prevent a similar accident in the future. It became one of their best practices and was then required at all plants. In this way, everyone can learn from the mistake of one of plant.

We also invited subject matter experts to these conferences. We invited a claims adjuster to talk about how injury claims were being managed for the company and what they needed from the safety and HR leaders. This helped safety champions learn about insurance operations and, at the same time, helped improve best practices to control claims. One of the claims leaders discussed the importance of reporting accidents early and how that helps the claim adjuster manage the claim and allows the injured person to recover faster. There was one very serious claim when, as a result of the early reporting requirement, the injured employee was released from the hospital the very same night. His family was assured that the expenses would be taken care of and he received the best care possible from the hospital knowing that the claim adjuster and the rehab nurse were available for any questions.

We invited a workers' compensation legal person to speak as well. He discussed the changes in workers' compensation laws and how they affect the handling of claims. He discussed the meaning of the clause "arising out of and in course of employment." The company's responsibility and legal requirement to prevent at-work injuries is based on whether the injury arises out of work and in the course of employment. Knowing the definition and context of the phrase "arising out of and in course of employment" is critical in the proper investigation of accidents and when taking corrective action. He discussed the discovery process and workers' compensation hearing process. And he took us through the claim and litigation process. When a claim is filed by an employee for an on-the-job injury, it is internally investigated and reported by the company to the workers' compensation insurance company. The insurance company claim adjuster then reviews the claim report and either authorizes payment or disputes the claim if he or she feels it does not arise out of or in the course of employment.

A letter is then sent to the employee stating the claim is being reviewed with the potential of being disputed. The adjuster can then authorize surveillance, which has to be done in such a manner that it is acceptable to the court under jurisdiction. The employee typically either accepts the claim adjuster's finding or dispute by filing a claim with the Industrial Commission. The company is notified by the Industrial Commission of the disputed claim. The claim can then be negotiated for settlement or litigated. Discussions like these helped our safety champions and HR leaders to better manage injury situations.

Workers' Compensation Claims Review

				Date Completed:	/10/04
Claim Office:		Policy Number:	WC	Claim Number:	9409
Policy Deductible Amount:	No	Expense Included in Deductible:	☐ Yes ☑ No	Date of Loss:	1/14/02
Insured:	Fort	Insured Location:		Date of Hire:	11-3-97
Claimant:	Ryan	Age:	27 yrs old	Occupation:	
Average Weekly Wage:	1,186.70	Weekly Comp Rate:	TIBS – $537.00 IIBS – $376.00	Monthly Rate:	n/a
Litigation:	☐ Yes ☑ No	☑ Initial Review ☐ Subsequent Review		Last Review Date:	n/a

☐ INITIAL REVIEW – Detailed description of accident (cause of loss, subrogation potential, theory of liability, and initial investigation):
☑ SUBSEQUENT REVIEW – Current status (investigation, litigation developments):

Employee was lifting a barrel of coating and his right shoulder popped.

☐ INITIAL REVIEW – Current status (diagnosis, prognosis, expected length of treatment/ disability):
☑ SUBSEQUENT REVIEW – Medical developments:

The employee sustained a compensable injury to the Upper back including cervical, and lower back. He remains currently off work. He had a Designated Doctor appointment on -20-04 with Dr. Milton for Maximum Medical Improvement and if so, what is the Impairment rating. He is receiving chiropractic treatment, to include traction and myotherapy. His diagnosis is 2mm bulge of his lower back and mild desiccation signal changes are present at his T3-4, T4-5, and T6-7 levels. He had undergone cervical facet injections and lumbar facet injections without any type of relief. The claimant is currently going to have a discogram to confirm the diagnosis. The surgeon wants claimant to have surgery. At this time, it is unclear as to what type of surgery, the surgeon would like claimant to undergo surgery.

EXPOSURE

ESTIMATION OF FINAL EXPOSURE AND JUSTIFICATION (Include plan for next 6 months. If closed, indicate date of closure and reason for closure.)

This injury is under New Law with lifetime medical benefits. The injured worker is currently seeking medical treatment and Impairment Income benefits. The injured worker has not been released to any type of work. I will continue to monitor this case for medical management and return to work full duty and/or start the process for a Designated doctor exam to evaluate Maximum Medical Improvement, and Impairment rating. Currently, we are assigning this claim to an Independent Medical exam for future medical treatment, and medications. We had a Peer Review that states that passive modalities should have been initially administered to the patient for about 2 to 3 weeks. Then the patient should have begun an active rehabilitation program to help strengthen and stabilize the areas of injury, as well as, good education. This should have lasted anywhere from 8 to 12 weeks seeing that there was 3 areas of injury which included the entire spine, of which a good home exercise program would have been given to follow with the patient.

CLAIM FINANCIAL HISTORY

Coverage Type	Outstanding Reserve	Paid Loss	Total	Expense Reserve	Paid Expenses
Indemnity	$11,366.29	$30,064.21	$41,430.50	None	None
Medical	$26,814.05	$43,736.26	$70,550.31	$2,255.72	$2,569.31
TOTAL	$38,180.34	$73,800.47	$111,980.81	$2,255.72	$2,569.31

COMMONWEALTH OF PENNSYLVANIA
DEPARTMENT OF LABOR AND INDUSTRY
BUREAU OF WORKERS' COMPENSATION
1171 S. CAMERON STREET, ROOM 103
HARRISBURG, PA 17104-2501
(TOLL FREE) 800-482-2383
TTY 800-362-4228
www.dli.state.pa.us

COMPROMISE AND RELEASE AGREEMENT BY STIPULATION PURSUANT TO SECTION 449 OF THE WORKERS' COMPENSATION ACT

Date of Injury: 06 / 07 /

PA BWC Claim Number:

Employee
First Name   Last Name
Street 1
Street
Street 2
City/Town   State   Zip Code
Philadelphia   PA   19135
County   Telephone
Philadelphia

Employer
Name
Street 1
Street 2
City/Town   State   Zip Code
King of Prussia   PA   19406
County
Montgomery
Telephone   FEIN
847-427-5301   0830

TO THE PARTIES: DO NOT SUBMIT THIS AGREEMENT TO THE BUREAU. SUBMIT IT TO THE ASSIGNED WORKERS' COMPENSATION JUDGE.

TO THE EXTENT THIS AGREEMENT REFERENCES AN INJURY FOR WHICH LIABILITY HAS NOT BEEN RECOGNIZED BY AGREEMENT OR BY ADJUDICATION, THE TERM "INJURY" AS USED IN THIS AGREEMENT SHALL MEAN "ALLEGED INJURY".

Insurer or Third Party Administrator (if self-insured)
Name
Street 1   P. O. Box 2060
Street 2
City/Town   State   Zip Code
Farmington Hills   MI   48333-2060
County   out of state
Telephone   Bureau Code
800-257-1900   0077
Insurer/TPA Claim Number   FEIN
1033572   210

1. This is an agreement in the case of the above listed employee and the above listed employer, insurer, or third party administrator in regards to an injury or occupational disease.

2. State the date of injury or occupational disease 06 / 07 /

3. State the average weekly wage of the employee, as calculated under Section 309. $980.73/wk

4. State the injury, the precise nature of the injury, and the nature of the disability, whether total or partial.

   Lower back and left lower extremity radicular symptoms as a result of the 06/07/05 work injury and all other sequela whether known or unknown, including but not limited to a specific loss.

5. State the weekly compensation rate paid or payable. $653.85/wk

6. State the amount of indemnity benefits paid or due and unpaid to the employee or dependent up to the date of the stipulation or agreement or death. $240,000.00

7. State the amount of the payment of indemnity benefits to be made at or after the date of the stipulation or agreement or death, and the length of time such payment of benefits is to continue. $240,000.00 for indemnity and medical benefits with medical benefits being left open until November 30, 2009.

LIBC-755 REV 4-04 (Page 1)

American Legalnet, Inc.
www.USCourtForms.com

On another occasion we invited the loss-control engineer from our insurance company to discuss loss trends. He explained that 69 percent of injuries are a result of strains and sprains and that we need to do something about ergonomics to take off some of the physical stress on our employees' bodies. As a result, we invited an ergonomist to the next conference, who discussed the importance of task analysis and developing engineering solutions for long term results.

We invited our insurance broker, Crissie Insurance Agency, to attend one of our monthly conferences. He discussed various important agency services that assist us in achieving world-class safety culture. He was able to help us select quality insurance carriers, Hartford and CNA Insurance Companies, for our insurance and risk management needs. He was able to get us online access to insurance company reports on any employee injury or property losses so that we could get our management and employees engaged immediately in mitigating such activities. His experienced staff were able to review our third-party contracts—temporary agency contracts, sub contract agreements, acquisition agreements, etc.—in order to improve our risk management and safety processes. The agency helped us with our claims review process so that injured employees could be treated fairly by the insurance company adjusters with an eye to actively managing claims with a team approach. And the agency invited us to various informative educational seminars on industry topics such as current insurance and safety issues. The agency truly acted as an important team member for actively engaging our employees and management.

We challenged members—safety champions and HR leaders—to present a topic of interest to them. Some of them had worked for fire departments and they were able to discuss the importance of emergency evacuation and CPR training. Ultimately, we used these monthly conferences to educate our safety champions and encourage plants to embrace best practices.

Safety Committee Meetings

Safety committees can be structured in several different ways; I have seen three types of safety committees:

1.   Management safety committee
2.   Management and employee safety committee
3.   Employee safety committee

At the corporate level, we created a management safety committee consisting of the VP of operations, HR director, and safety director. We met quarterly and set the tone for the safety program throughout the corporation. If there was a need to change direction or introduce a new program, we would discuss it and make a judgment at our meetings.

At every plant, other than those in Canada, we had joint management and employee safety committees. Typically, there were six to eight people on the company safety committee because I worried that large committees might not be able to accomplish meaningful goals. For a safety committee to be effective it has to be fully functional. Each meeting has to have a safety plan and an agenda. These needs to be followed through and monitored toward success. The purpose of these meetings is to establish a world-class safety culture, limit the number of employee injuries, and make sure every employee returns to his family safe and sound.

One of the requirements of the plant safety committee was to complete monthly audits using a checklist of the facility. Later on we added requirement that those completing audit also needed to observe employees at work and comment on any unsafe behaviors. The plant safety champion was the chair of the safety committee and he or she was required to distribute the minutes to the members and general manager within a week after the meeting. The minutes were also posted on the safety board for employee review. Any recommendations that came as a result of the safety audit were then reviewed at the next meeting and a determination was made as to the status of the recommendation in compliance. Not only we were able to get employees involved in the process of safety this way, but we were also able to develop a good relationship with management and employees as a result of working together to cultivate a safe workplace.

Our Canadian operation in the province of Ontario was a little different. Different from OSHA standards, in Ontario the Canada Labour Code (Occupational Health and Safety Act (R.S.O. 1990, c. 0.1, Section 9(2)) requires all companies to have a joint health and safety committee without the active participation of management; it had to be run by an employee representative. This was a mandatory requirement. The committee functioned the same way as described above except that it had to be run by an employee representative as the chairman. But before assuming the responsibility as a chairman, the employee had to undergo an intensive two-day training on safety and how to run the safety committee. He or she was then certified and authorized to operate the safety committee.

Human Resources and Skills Development Canada
Fair, safe, and productive workplaces

Information on Occupational Health and Safety
6B Workplace Health and Safety Committees

Introduction

The *Canada Labour Code* protects the rights of employers and employees and establishes a framework for the resolution of disputes. The objective of Part II of the Code is to reduce, as much as possible, the number of employees who suffer casualties as a result of their work activities.

This pamphlet explains the Code's requirements regarding **workplace health and safety committees**, to address health and safety issues.

1. **To whom does the requirement apply?**

The requirement applies to every employer in the federal jurisdiction. A workplace health and safety committee must be established for each workplace, controlled by the employer, that has 20 or more employees.

2. **Who sits on the workplace health and safety committee?**

A workplace health and safety committee consists of at least two persons. They are appointed by the employer, in accordance with the following conditions: at least half of the committee members are employees who do not exercise managerial functions. These members are selected by the trade union representing the employees in consultation with any employees who are not so represented. If they are not members of a union, then employees at large will select their committee representatives.

If the organization does not have a policy committee, then a work place health and safety committee, when dealing with an issue that would have gone to the policy committee, can select two additional members. One of those additional members is selected by the trade union or by the employees at large.

A work place health and safety committee is led by two chairpersons, one of whom is chosen by the employer-members and the other by the employee-members.

Terms of office are not to exceed two years.

3. **What are the powers and duties of a work place health and safety committee?**

There are several. The work place health and safety committee will:

- consider and expeditiously dispose of health and safety complaints;
- participate in the development, implementation and monitoring of programs to prevent work place hazards, including ergonomic related hazards, if there is no policy committee in the organization;
- participate in all inquiries, investigations, studies, and inspections pertaining to the health and safety of employees;
- participate in the implementation and monitoring of a program for the provision of personal protective equipment, clothing, devices, or materials, and, if there is no policy committee, participate in the development of the program;
- ensure to keep adequate records of work accidents, injuries, health hazards, health and safety complaints and regularly monitor this data;
- cooperate with health and safety officers;
- participate in the implementation of changes that may affect occupational health and safety, including work processes and procedures, and, if there is no policy committee, participate in the planning of the implementation of those changes;
- assist the employer in investigating and assessing the exposure of employees to hazardous substances;
- inspect each month all or part of the work place, so that every part of the work place is inspected at least once a year;
- participate in the development of health and safety policies and programs, if there is no policy committee.
- participate in the development, implementation and monitoring of a work place violence prevention policy, if there is no policy committee in the organization.

The committee may request from an employer any information that it considers necessary to identify existing or potential hazards in the work place. It has full access to all government and employer reports, studies and tests relating to the health and safety of employees. Of course, the committee does not have access to the medical records of any individual except with the person's consent.

4. **Do the members of the work place health and safety committee receive training?**

   Yes. The Code requires the employer to ensure that committee members receive the necessary training in health and safety and are informed of their responsibilities under Part II of the Code.

5. **Are there exemptions from the requirements?**

   The Minister of Labour may exempt an employer from the requirement to establish a work place health and safety committee if the Minister receives a request to do so and is satisfied that the work place is relatively free from risks to health and safety. The factors that the Minister must consider in determining whether to exempt an employer are listed in the Code. Those factors include the risk of injury or illness from hazardous substances in the work place; the physical and organizational structure of the work place and the types of work being done; the nature of the operation, the work processes and the equipment used; the number of disabling injuries in the work place in relation to the number of hours worked in the work place; the occurrence of incidents having serious effects on health and safety; and any contraventions of Part II of the Code.

   When exempted from the requirement, the employer must appoint a health and safety representative for that work place.

6. **Is the employer required to compensate the members?**

   Yes. The employer must pay committee members at their regular rate of pay or premium rate of pay, as specified in the collective agreement or, if there is no collective agreement, in accordance with the employer's policy.

   The requirement to compensate members applies to:

- attending meetings or performing any of their other functions; and
- preparation time and traveling as authorized by both chairpersons of the committee.

7. **How often will the committees meet?**

Each work place health and safety committee is required to meet 9 times a year, at regular intervals and during regular working hours. If circumstances make additional meetings necessary, they should be held during or outside regular hours, whatever is required.

The employer must ensure the availability in the work place of premises, equipment and personnel for the efficient operation of the committee.

8. **What other administrative issues do I need to know about?**

First, with respect to rule-making, the work place health and safety committee can establish its own rules of procedure pertaining to the administration and operation of the committee (e.g., meeting times and places).

Secondly, with respect to record-keeping, the committee must keep accurate records of all matters that come before it, as well as minutes of all meetings. They are to be made available to a health and safety officer if they are requested.

Thirdly, with respect to liability, no committee member is personally liable for anything done, or not done, in good faith under the authority of the committee.

Finally, with respect to regulations, the Minister of Labour retains the right to make certain regulations which may apply to all of the work place health and safety committees, to a group of committees, or to a single committee. For instance, the Minister may specify the method of selecting members of a committee if the employees are not represented by a trade union. The Minister may make a regulation specifying the manner in which a committee may exercise its powers and perform its functions.

It does not matter how the safety committee is structured. The important thing is to make sure the committee is effective. As long as there is rapport and understanding among the committee members and management, along with a solid mission of improving safety, everything should fall into place.

To be effective in providing a safe place to work, the safety committee should do the following:

- Define clear goals and expectations
- Create an agenda
- Measure the committee's progress
- Be proactive
- Be passionate about the safety of employees
- Maintain clear, concise, and proper communication between the committee members and management
- Follow assignments through to execution
- Run committee meetings as if they are business meetings—to discuss the safety of business

Safety committees can be very effective in making sure employees are safe. However, they have to be run efficiently to be impactful and accomplish their vision and mission for safety.

Engage employees using the power of positive thinking. I was forced out of my previous company and my initial reaction was to be upset with the company for not giving me the benefit of doubt. This lasted for a couple of months and then I told my brain, "It was meant to be. It is God's wish." Once I was able to get past the immediate negative reaction, I was able to think about what I could do with my knowledge and experience. The first thing I did was try to write an article on what I knew well: safety and risk management. I wrote an article and sent it in for publication. It was rejected. I went back and improved the writing based on the feedback I got. The third time it was accepted for publication. This was an exciting development because it allowed me to show my talents in a different way and help professionals in the field.

As an offshoot of that positive outcome, I decided to write this book. So I would like to thank Shirzad Chamine for an outstanding thought in his presentation on positive intelligence at the ASSE 2015, ASSE's premier Professional Development Conference in 2015. As he stated, the DNA of an organization is the combined sum of the DNA of the individuals who work there. There are those in an organization who are highly positive, there are those who are mildly positive, and then there are those who are negative. The organization's reaction to an event will be determined by the combined sum of the individual positive or negative intelligence determined by the Positive Quotient (PQ). If you cultivate positive PQ in individuals, you will find that the organization as a whole will react more positively since more employees have taken to be thinking positively.

Engage employees in positive thinking and positive safety work, and you will get results as shown by some of the comments we received from our employees after we created a safer environment:

"To me, safety has become a serious subject and I now understand the success in developing a safe environment is dependent on individual responsibility and participation by all."

"In my twenty-five years in this industry, I have never experienced the emphasis placed on safety as I have here."

"Working here has greatly affected my view of safety both at work and at home."

"Safety has become second nature to me now that we have received all of the training here at work."

These comments are a result of positive feedback and rewarding people for positive behaviors. At the same time, success breeds success. People tend to jump on the bandwagon when things are going in a positive direction.

We look out for each other and this is encouraged throughout the corporation. Looking out for each other is paramount in a company with world-class safety culture or that is striving to be world class. There is an interesting book called *Would You Watch Out for My Safety?* by John Drebinger on this subject. Drebinger has made several presentations at professional development conferences on this subject. This custom should not only be preached regularly, but also practiced and publicized locally in company newsletters and memos, where it becomes contagious.

As shown in the example below, when John saw Steve moving a drum off of a pallet, he realized the potential danger of Steve moving the drum by himself. He pitched in and helped Steve move the drum safely. In this particular case, the incident was publicized by posting a flyer. The company took the extra step and purchased an electric drum handler at a significant cost in order to reduce the likelihood of a severe back injury for any employees.

Material Handling Process
Catch Associates Doing Right

- ☐ Steve was moving the drum on a pallet.
- ☐ John saw he needed help to prevent a strain.
- ☐ He pitched in to help.
- ☐ John was recognized for seeing the hazard and the need to help a fellow associate.

PROCESS

You can get employees actively engaged if you touch their hearts and ask them to do something they enjoy doing. Every one of us has talents other than the tasks we do at work. The challenge is to identify employees who have these talents and then find ways to tap into their skills to achieve safety-related goals.

At FDC, it was our goal to bring people back to work as soon as practicable after they were injured so that they could feel productive and also help improve their personal morale. One employee, a feeder at a Mitsubishi press, injured his back while feeding paper into the press. He needed back surgery and the doctors recommended a couple of months off work for recovery. He was talented with computers and had been involved in teaching computer skills at the local community college. We tapped into his talents and got him involved in teaching computer skills to our employees. He was happy that his talents were being recognized and that he was being gainfully employed. The other employees who he taught were happy to learn and realized that the company cared enough to find a way for an injured employee to continue to contribute to the company rather than dwell on what he is no longer able to do. It was a win-win situation and a positive experience for all involved. I am not saying every injured employee can be so gainfully employed. I am saying that taking a little more interest in your employees can help you find a way to bring an injured employee back into a gainful employment.

Identifying employee talents and finding ways to use them can be helpful in getting employees engaged in safety. We had published a corporate safety newsletter that I used to send out regularly. However, it did not have the local flavor, so I was told by an employee in the finishing department at the South Carolina plant. I asked him if he would like to put together a page or two for the plant safety newsletter that would include more local safety news and flavor? He took on the challenge and put together a quarterly newsletter that contained the safety training information offered by the company, safety news reported in the local newspapers, newspaper articles on accidents and injuries to locals, and any other safety-related communication. The plant employees loved the local perspective and before long every plant started putting out its own local safety newsletters.

Another interesting way to actively engage employees is holding a safety contest. We asked our employees to respond to the prompt, "What Safety Means to Me." The participation in the safety contest was overwhelming with great responses. These were the winning entries:

- Safety is contagious; it spreads.
- Safety is important to me. When we are safe, people are healthy; when people are healthy, they are productive and effective; when people are productive and effective, they are better at their jobs; when people are better at their jobs, they are able to make the right decisions, make good products, and make customers happy; when customers are happy, they buy more products and tell their peers about the good experience they have had with the company; when customers buy more and tell others about the company, we sell more, make more, and earn more; when the company sells more, makes more, and earns more—our company is more successful. In conclusion, when we are safe—we are more successful and that is not just a good thing, it is a GREAT thing!!
- The Safety Circle
  - Safety training = confidence and trust in practices
  - Confidence and trust = better efficiencies/productivity
  - Better productivity = more time to dedicate to safety training and $$$

The CEO congratulated the winners and the participants. All participating entries were placed on the corporate safety board for everyone to see and appreciate.

Another interesting and rewarding program is a game called "What is in this picture?" For this game I took pictures of various employees at work or in unsafe conditions and placed the pictures on the safety board, asking employees to identify as many safety hazards—unsafe conditions and unsafe acts—as they could. Employees placed completed papers in a box and at the end of the week the HR manager reviewed all of the answers and selected the one with the most hazardous conditions or hazardous behaviors identified. The winner would receive a gift certificate for $20.

Not only did employees get engaged in identifying hazards, but they also started learning about safety in the process and taking safety procedures at work more seriously. It created an exciting way to engage employees and improve the safety process.

A Key Performance Indicator (KPI) is a business metric used to evaluate factors that are crucial to the success of an organization. KPIs differ by organization. Every week the VP of operations at our company would start the meeting that all key executives, including the general managers of each plant and directors of various corporate departments, attend by reviewing the previous week's KPIs, which included safety, quality, productivity, sales, and customer complaints. By doing so he was able to get key employees engaged in the proper operation of the business. Safety process became one of the key components of the business operation. I had to discuss if there were any accidents and what actions were being taken to prevent them, frequency and severity rates as of that week, and the cost of injuries for the week and year to date.

The program required the interaction of all the key players in the organization and provided an avenue for keeping each other aware of various business trends and what was happening in the business as of that week. Attached is a safety chart showing the incidence rates compared to the industry average and the cost of injuries, which I prepared for the monthly CEO meetings.

ABC Company Chart
LWDII Chart

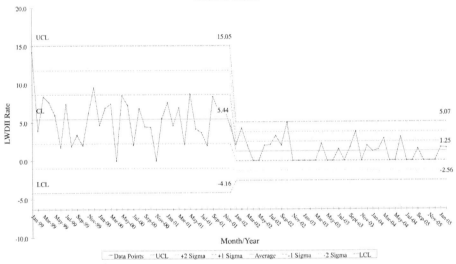

PROCESS

The company offered each employee a bonus at the end of the calendar year based on various factors. These could change based on the direction the CEO wanted the corporation to go in. The list on the next page shows the criteria he used to determine bonuses one year. As you can see, safety was a key component in the employee bonus determination.

Bonus Plan
Performance Metrics
- Safety
- Sales
- Operating Profit
- OTIF

Again, the criteria he established encouraged employees to be engaged in safety so as to improve their chances of receiving a larger bonus.

Employee engagement surveys are critical in gauging employee feelings and learning how employees would react in a safety situation. When I was hired at FDC, it was a family owned operation and the owners offered overtime during Christmas holidays even though we had no work. They wanted employees to have a nice holiday season with their families. The company was not making enough profits and the bank executives were at its corporate door every week. The reason, I believe, was the fact that the company was now owned by several children of the original owners—the family had been running the company for three generations—each child had his or her own demands on how the company's finances should be managed. As I mentioned in Chapter 1, the owners finally decided to sell the company to a private equity firm because it could not satisfy the demands of each of the siblings. The new board of directors brought in a new executive team who believed in a different culture—their shareholders demanded increased rates of return. The executive team had to change the culture of the organization to more of a productive environment. The new team asked Gallup to complete a survey of the employees so that they could accurately evaluate company culture.

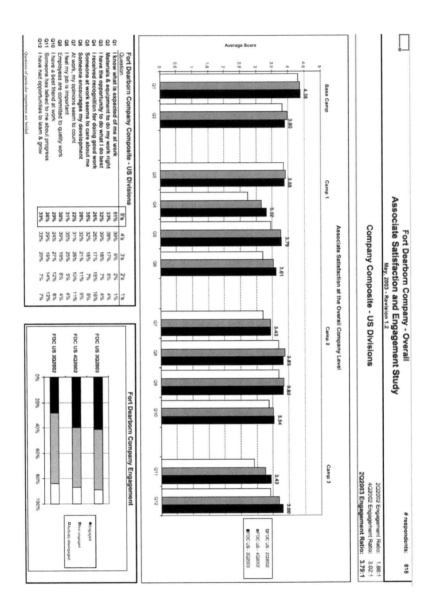

Powered with that tool, the executives put together a new company vision. Culture creates the climate necessary for change. The solutions come from executives and people in trenches. The new CEO put together his vision and defined where he wanted the company to go. Posters with the new vision and core values were then placed all over the corporate offices and plants.

As regards safety, twice a year each plant independently surveys its associates to evaluate how engaged they are in the success of plant safety. In addition, once a year the corporate safety department completes a survey of how the associates are engaged in safety. We devised a checklist of twenty-five safety-related questions and provided a copy of the survey with envelops at random to about 12 percent of the total division population. The associates' responses are mailed to the corporate safety department where these are tabulated and scored on a scale of one to five, one being strongly disagree and five being strongly agree. The following are the results from the last survey:

Fountain Inn – 86%

Niles – 82%

Lake Forest – 77%

Fort Worth – 74%

Flexible Packaging – 74%

Virtual Color – 72%

The individual line-item scores and total-point scores and comments were reviewed with each general manager for improving associate engagement score. See the below scoring card and rating sheet for your review and use.

Location: Plant 3						# of people who took survey: 19	
Questions (1=Strongly disagree, 5=Strongly Agree)	1	2	3	4	5	Total	Score
1. Management operates an open door policy on safety issues.	0	0	2	6	11	19	89%
2. Safety is a priority in my mind when completing a job.	0	1	3	5	10	19	79%
3. Co-workers often give tips to each other on how to work safely.	1	2	3	5	8	19	68%
4. Safety rules and procedures are carefully followed.	1	1	2	8	7	19	79%
5. Management clearly considers the safety of employees of great importance.	0	1	2	7	9	19	84%

Location: Plant 3						# of people who took survey: 19	
Questions (1=Strongly disagree, 5=Strongly Agree)	1	2	3	4	5	Total	Score
6. I am given enough time to get the job done safely.	1	2	1	7	8	19	79%
7. I am involved in informing management of important safety issues.	1	2	1	8	7	19	79%
8. Management acts properly when a safety concern is raised.	2	0	4	5	8	19	68%
9. There is a good communication here about safety issues which affect me.	0	2	2	8	7	19	79%
10. I understand the safety rules of the job.	0	1	0	4	14	19	95%
11. It is important to me that there is a continuing emphasis on safety.	0	1	1	2	15	19	89%
12. This is a safer place to work than other companies I have worked for.	1	1	5	5	7	19	63%
13. I am strongly encouraged to report unsafe conditions.	1	1	2	6	9	19	79%
14. Personally I feel that safety issues are not the most important aspect of my job.	7	6	2	2	2	19	68%
15. In my workplace the chances of being involved in an accident are quite large.	8	6	3	1	1	19	74%
16. I do not receive praise for working safely.	7	7	2	2	1	19	74%
17. I can influence health and safety performance here.	1	2	2	6	8	19	74%
18. When people ignore safety procedures here, I feel that it is none of my business.	8	6	2	3	0	19	74%
19. I am clear about what my responsibilities are for health and safety.	0	0	2	5	12	19	89%
20. A safe place to work has a lot of personal meaning to me.	1	0	0	5	13	19	95%
21. There are always enough people available to get the job done safely.	1	2	3	7	6	19	68%
22. In my workplace managers/supervisors show interest in my safety.	1	2	2	6	8	19	74%
23. Management considers safety to be equally as important as production.	1	2	3	7	6	19	68%
24. Managers and supervisors express concern if safety procedures are not adhered to.	2	1	3	7	6	19	68%

Location: Plant 3						# of people who took survey: 19	
Questions (1=Strongly disagree, 5=Strongly Agree)	1	2	3	4	5	Total	Score
25. I cannot always get the equipment I need to do the job safely.	7	6	3	2	1	19	68%
							77%

Notes: Questions are designed to force employees to read and answer them honestly. For scoring: Add the last two columns of all questions except 14, 15, 16, 18 & 25, for which you add the first two columns and then divide by 19 participants for percent scoring.

It is extremely important to gauge the employee pulse with a meaningful survey of employee feelings so that improvements can be made at each plant and resources can be provided where meaningful progress is needed. Identifying where improvements can be made will increase employee buy-in to safety initiatives.

World-class safety companies have outstanding collaboration between safety team leaders and human resources personnel both at the corporate level and at each of the plants. Each location has its own responsibilities, which sometimes overlap, but in every case, both the safety team leader and human resources manager need to be in constant communication about fulfilling their responsibility to management and employees.

At FDC, corporate HR and the safety director worked together to identify the plant of the year, determine the language to be used in return to work programs so that they were not discriminatory, and develop the employee handbook and employee safety rules. At the plant level, HR managers and safety team leaders worked together to run plant safety committee meetings and organize employee engagement activities such as safety training, self-inspections, job safety observations, and safety incentives.

As corporate safety director, I realized long ago that if HR managers and safety team leaders do not actively work together, it is difficult to achieve a world-class safety culture. Every year when we had our safety summit, the company would invite HR and safety team leaders to a day and half seminar at the corporate offices so that the two groups could get to know each other better and also come to know their counterparts at other facilities.

A typical agenda included an introduction from the CEO, a report on the state of the company's safety initiatives and future plans from the corporate safety director, a half-day presentation by an invited speaker, followed by presentations from each plant on the status of their safety—losses and injuries, current safety initiatives, and future safety plans. At one of the safety summits, I took all the leaders to a one-day seminar at the Chicagoland Safety Conference in Naperville. Another time we held a round table discussion on current safety topics. Some of the comments I received from the summit were:

- I liked travelling to Elk Grove and meeting safety team leaders from our other plants. I really enjoyed this meeting. The guest speaker was really good; his presentation was well done and highly informative
- I liked the plant presentations the most. It was a great platform for ideas and brainstorming. I also enjoyed the speaker. His analogies helped me understand the point he was driving—how to change the safety culture.
- I liked the new ways to address safety by addressing safety culture.
- I liked the discussion on merging safety performance and production performance into an overall operational performance system. If there is an accident, it's a system failure which should be treated no differently than a quality reject or a label mix.
- To make lasting change in your facility with regard to safety, you have to change the culture and that starts from the top down.
- We can have an influence on others in the work place even though we may not realize it at the time. All managers need to practice or demonstrate a positive image especially when it concerns safety.
- I plan to survey associates to determine where they rank safety

in their job satisfaction in their list of top ten job satisfiers so that I know how high safety is in their list of job satisfiers.

- I liked the team building activity. It was very interesting.
- This summit gave me an opportunity to get to know my counterparts. I learned various insights from the safety conference I attended.

What I am trying to emphasize is that an annual safety summit consisting of HR and safety team leaders has a positive impact on learning and developing rapport between these two important departments. World-class companies get everyone engaged in safety. For instance, we got an insurance company consultant engaged in completing accident analysis and safety audits. He also attended our safety summit, willingly participated in safety discussions, and even made a safety presentation. His accident analysis using the secured loss analysis system demonstrated the following:

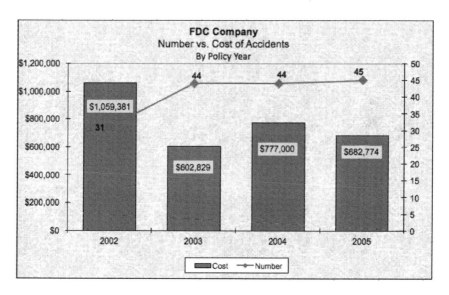

The graph shows that the cost of injuries decreased from 2002 to 2005 from $1,059,381 to $682,774 even though the total number of injuries for 1500 employees increased from 31 to 45. The bulk of these injuries were minor in nature but still needed to be controlled. Getting professionals from the insurance companies involved in this effort helped the company achieve its goal of developing a world-class safety culture. During one of his safety audits, the Hartford Insurance Company consultant noted:

> "Although there was a very strong commitment and formal program in place, the company was eager to reach out and learn in an effort to continually strengthen their program. Secondly, we quickly realized that this program had the full support of management. A management influence is present in all safety activities, and there is no doubt that safety is a key component of their business."

Another way that we engaged employees in safety was by requesting their participation in special projects. We put together sub-committees as needed within the realm of the safety committee. The insurance company auditors and state OSHA auditors requested that we put a comprehensive Lock Out Tag Out (LOTO) program in place and institute LOTO procedures for all major pieces of equipment. We formed a sub-committee for this purpose, and they were able to put together a program and procedures like the example shown below.

LOTO Procedure

PURPOSE: To provide a system that will ensure the safety of all associates in the Fort Worth facility.

SCOPE: Authorized associates performing any task on banding machine that requires LOTO.

RESPONSIBILITIES: Authorized and trained associates within the plant. In the event that an associate who has locked the equipment has left the plant and is not available by phone to return to the plant, a duplicate key is available to change the lock. However, to remove the existing lock without replacing it with another lock requires the approval of the maintenance department. The maintenance department must complete an inspection of the machine to insure there are no safety risks in removing the LOTO. Duplicate keys are accessible to managers only.

DEFINITION:
- Authorized Associate – Are those who perform minor servicing tasks or maintenance to equipment and are required to LOTO or use alternative protection measures.
- Affected Associates – Are those who work in an area where minor servicing or maintenance is being performed, and are required to understand the purpose of LOTO.

ENERGY SOURCE: Electric energy and pneumatic energy

LOCK TYPE: Padlock and valve lock

DE-ENERGIZE: Electric and pneumatic energy

SHUT DOWN PROCEDURE:
1. Push emergency stop button.
2. Turn off the main circuit switch, insert padlock and tag.
3. Turn off the air valve disconnect, relieve pressure, apply valve lock and tag.
4. Keep the key with you to avoid the machine being unlocked by associates who are unaware.
5. If more than one associate is performing the task, each associate must have their own lock and tag applied.

VERIFY:
1. Push the start button to verify it will not start.
2. Perform maintenance required.

STARTUP PROCEDURE:

1. Verify all guards are put back in place and all associates are clear from equipment.
2. Unlock and remove lock, tag, etc.
3. Verify equipment is operational.
4. Notify manager, department lead, etc., that task is complete and ready for production.

Banding machine

Main circuit switch

Air valve disconnect

We were not only able to satisfy the insurance company and OSHA, but we also got the workforce excited about being a part of something good and rewarding.

In a world-class company, safety is always included in measuring company success and planning process. The company I worked for had a program called "Snapshot." Every leader had to prepare a monthly snapshot of where they are and where they are heading using the following captions: Goals and Objectives, Current Snapshot, Performance, Green/Red Flags and Action items (an example of this can be seen in Chapter 2). This program has significant impact in getting management engaged.

We feel that safety is everyone's responsibility and want to include everyone, including shareholders, executives, managers, and union and nonunion employees. As such, we developed a progress report and attached it to every payroll check defining the role of each associate. This helped every associate get actively engaged.

As the chief operating officer of the corporation, the role and responsibility of the **President** is to furnish each associate with a place of employment that is free from recognized hazards that could cause injury of any type, serious or minor, to our associates and visitors.

The role and responsibility of the **Corporate Safety Director,** who is in charge of overall company safety-related goals and objectives, is to:

a. Plan and implement company safety policies and procedures to comply with government rules and regulations.
b. Coordinate company-wide programs to ensure work site safety practices.
c. Audit FDC facilities to detect existing and potential accident and health hazards and offer advice to prevent these hazards.
d. Assist in the investigation of lost time accidents.
e. Analyze accidents and develop trends for the purpose of reporting corrective action.
f. Support safety team leaders to achieve company safety goals.
g. Coordinate the development of programs and procedures to ensure safe behavior.

With regard to safety, the role and responsibility of each **General Manager** is to support, encourage, and lead the safety team leader, department managers, and associates in fulfilling the Safety AIM or the desired state of the company.

The role of the division **Safety Team Leader,** who also has other management responsibilities, is to:

a. Complete required safety training of associates and maintain records of the training.
b. Chair the safety committee meetings.
c. Complete semi-monthly safety inspections using the safety checklist.
d. Investigate all accidents and complete a report of the investigation with corrective action.
e. Motivate associates to work safely.

The role of each **Department Manager** in the division is to:

a. Complete weekly tool box meetings.
b. Complete a daily safety inspection of the department.
c. Make suggestions on how to improve the system.
d. Complete a daily safety huddle.
e. Assist in the investigation of accidents by Safety Team Leader and Safety Director.
f. Train associates in safe work practices.
g. Observe associates working safely.
h. Coach and counsel associates to ensure safe behavior.

The role of each **Associate** (employee) is to ensure a safe work place by working together with managers and division safety team leaders and includes the responsibility to:

a. Obey all safety rules and regulations.
b. Participate in the training meetings.
c. Make suggestions on how to improve the system.
d. Work safely as an individual and as a team.
e. Ensure the work area is kept clean and free of hazards.
f. Report all unsafe acts and conditions to the immediate supervisor.

We discussed the role and responsibilities at various meetings held at the corporate office and at other divisions so that senior executives, managers, and associates were aware of what was expected of them. At safety team leader meetings, we discussed the importance of developing division safety committees and ensuring that committee members complete all desired safety activities. At the manager meetings, we discussed the importance of making sure all managers are trained in root-cause analysis, accident investigations, and the 10-hour OSHA certification program. When meeting with associates, we made sure they understood the safety rules and discussed how the safety observation program would be used to comply with the rules. We provided gift certificates and safety products as awards to associates for safety observations and hazard prevention ideas. For general managers and department managers the rewards for actively engaging in safety were determined in a different way. Annual bonuses for these managers were determined only based on corporate and plant financial performance. The CEO changed that. Per his instructions, 25 percent of manager bonuses were determined by safety-related accomplishments—reduction in LWDI and reduction in the incurred cost of accidents versus the expected average cost. The annual bonus could go up or down by this percentage based on safety-related accomplishments.

We met with union leaders and got them to buy into our safety initiatives. We met with the executives and changed how we charged each plant with the cost of workers' compensation insurance. Previously, each plant paid corporate a contribution based on workers' compensation payroll. As a result, the general managers did not have any real incentive to manage the cost of workers' compensation insurance. We changed that. We purchased a large deductible insurance program, which required us to pay a fee for insurance service and pay the workers' compensation cost incurred by the insurance company within thirty days of payment.

We made each plant responsible for its own workers' compensation costs. At the end of the year, each plant was charged back a percentage of the service fee based on payroll and the total cost incurred for all injuries at that plant for the policy year. As a result, the plant general managers watched for all claims and made sure conditions or processes causing the injuries were properly managed. At each of our facilities, a manager, instead of a volunteer associate, was designated as the safety team leader who would manage the safety program.

One of the most important responsibilities of the safety team leaders is to complete a monthly trend report of employee injuries for each of the divisions they represent, including injury IR and LWDII rates. When everyone gets involved, accidents are effectively controlled because everyone wants to prevent the cause of accidents and injuries.

Systemic Integration of SH&E and Business Functions

Systematically integrating safety, health, and business functions can be very rewarding. When we are looking at a safety function, we are also thinking about the overall safety system—the system being used to identify and control hazards that are causing harm or have the potential to cause harm to employees, customers and company property. This includes thinking about the safety and health inspection program, the accident investigation program, the employee training program, and other programs and initiatives geared to protect employees and manufacturing operations. When we think about business functions, we are thinking about how we are going to manufacture the products or services, finance the operations that manufacture the products and services, and effectively market the products and services to customers.

FDC had sophisticated safety and business functions that needed to be integrated for outstanding safety results. Although the financial results from business operations were good, the safety results needed improvement. Their accounting books had to be audited annually so that stakeholders would know how the company was performing financially. ISO auditors would check the quality control and operational systems to make sure these systems were on par with industry norms. In the same token, we had to establish a system that required outside auditors from an independent insurance company to come and audit our safety systems. They made their own recommendations for continuous improvements.

I believe safety and health functions should be blended properly with the corporate structure and that the Environmental Health and Safety (EHS) Director should be constantly involved in the company's operations.[9] The EHS Director for world-class companies is regularly involved in senior executive meetings, which include meetings to decide business strategies such as acquisitions or the sale of assets. We will discuss how this systemic integration occurs through the following points:

- A direct reporting relationship with CEO or COO
- Financial budget
- Monitoring of safety goals and strategies
- Aligning Plant of the Year Award with the Safest Plant of the Year Award

A direct reporting relationship with the CEO or COO is a manifestation of senior management's understanding of the importance of safety for the success of the business. It is an example walking the walk and talking the talk; it is demonstrating to every associate in the organization that safety is a critical business function. I was involved with FDC when it was in the process of acquiring a new plant in Brunswick, located on the eastern coast of Georgia that has significant exposure to hurricanes. The building was not designed for this kind of exposure. So we negotiated that the sale of the

---

9   Peter R. Scholtes and Russell L Ackoff, *The Leader's Handbook: Making Things Happen, Getting Things Done* (New York: McGraw-Hill, 1997).

building include a significant amount of improvement to protect the building and ensure that it could sustain higher "100 Year" wind exposure.

In another situation, the company acquired a large printing press with the intent to move the press to one of its plants in three months. During this three-month period, the printer was going to be operated and maintained by the seller. The safety director intervened and required that the seller provide certificates of insurance naming the new company as also insured while the press was in its care and custody. A week after the closing there was a serious accident on the press and this intervention by the safety director saved the company a significant amount of money and spared its executives some tough questions from the board of directors. A direct reporting relationship, reporting directly to the CEO or COO of a company, has to be earned by safety professionals.

In this case the buyer required the seller to include a clause in the purchase agreement to protect the buyer in case an incident at the press occurred. The clause also made sure there was proof of insurance through a COI that protected the buyer. And when that incident happened, the buyer was protected by the selling company and its insurance carrier.

By the same token, when you hire employees from an agency or independently, it is important that there is a contract and workers' compensation COI that will protect your company.

A Certificate of Insurance (COI) is "a document acknowledging an insurance policy has been written setting forth in general terms what the policy covers." Typically, the COI is a snapshot of basic policy coverages and limits at the time of issuance of the certificate. Certificates are not intended to modify coverages or change the terms of the insurance contract which they certify. COIs are provided either by the insurance company who issued the policy or by an insurance agency who represents the insurance company and issues the certificate on behalf of the insurance company.

Most insurance certificates are created to provide insurance policy information for interested third parties. They may be produced as a requirement of a contract between the named insured on the policy and the third party involved. A COI typically proves that a certificate holder's workers' compensation insurance policy exists. Please note that in most states, workers' compensation coverage is either in place or it is not. There isn't a limit or amount of workers' compensation insurance provided on the COI—all benefits are covered. A COI may also convey information to the certificate holder as required under their contract with the named insured if the certificate holder as shown is also an additional insured under the named insured's policy for that specific job or location. The extended insurance coverage is limited to that specific job or location.

For workers' compensation coverage for staffing companies and the employers utilizing them, it is important to note the rule on liability and workers' compensation insurance coverage for work-related injuries is not clearly known by even some veteran risk managers. The law in Illinois requires that both companies are jointly liable for a work-related injury to a staffer. In short, both companies are on the hook. Sometimes there is a battle over primary liability. The answer to the question of primary liability is simple: the company where the injured person was working is primarily liable unless there is an agreement to the contrary. Please ensure that you have a written agreement as to who has the primary liability.

Therefore, when you retain a staffing company, be certain to ask for documentation confirming they are not only providing you with workers, but that they are also providing primary workers' compensation coverage for any injuries suffered by the workers. Remember, if you are told they have workers' compensation coverage, that simply means what it says—they have workers' compensation coverage, which is required by law. It does not necessarily mean they are providing primary coverage of injuries. This is one situation where you want to look at the fine print to ensure you are clear about who may have primary coverage. The difference can cost thousands of dollars.

Another question that always arises is, "Should I let a sole proprietor opt out of workers' compensation coverage and still work for me?" The Illinois Workers' Compensation Act and laws in various other states allow a sole proprietor or officers of a company to opt out of coverage and save the premiums associated with workers' compensation insurance for themselves. Please remember that if you let such an individual onto your worksite or facility, you take a giant risk. If that man or woman is killed as the result of a work-related injury, the minimum death benefit in Illinois, for example, is currently about $625,000. The maximum death benefit in Illinois is over $1.6 million dollars. It is also possible for a worker to require lots of medical care and potentially need Temporary Total Disability (TTD) before passing away. Most widows or widowers in such a situation are willing to take a shot at making a claim even if their spouse didn't insure for such risks. Widows and widowers make sympathetic claimants.

In short, the exposure that comes from allowing a potentially uninsured contractor on your job site is massive. For a small or mid-sized company, this could be a business-busting liability. I strongly recommend against letting anyone who opts out of workers' compensation coverage to work on your company's premises. Make sure they have coverage for themselves and all their workers at their cost and not yours.

You also must realize that there are limitations to workers' compensation certificates of insurance. It is simply a document that provides information about workers' compensation insurance policy coverage. In today's world, a COI is used every day to provide valuable policy information to third parties. Generally, a COI is used when someone requests to be supplied with proof that the named insurance policy carries workers' compensation insurance, which is usually required by a contract or continuing business relationship.

The COI is designed to provide rudimentary policy information. Remember, it is workers' compensation fraud in most states to provide a false COI. While we must understand what the certificate is, we are also repeatedly asked what a COI doesn't do.

A COI cannot change a workers' compensation insurance policy; it is only an informational document and does not guarantee or preclude changes or endorsements to the insurance policy; it doesn't extend insurance policy conditions to the certificate holder; it can't modify the terms within the WC insurance policy; it doesn't guarantee an insurance policy will not be cancelled in accordance with the conditions of the policy; it doesn't bestow or extend new rights to the certificate holder; and it doesn't definitively provide full insurance coverage to the certificate holder.

There are several things you should look for in a COI, which are crucial to the world of business. As an insurance document, the COI provides information about insurance policies and coverage restrictions to interested third parties. Most of those third parties are involved because of some type of contract. Most often, certificates are asked for when there is some type of contract involving two or more companies.

Here's a list of the type of information you can find on a COI:

- Name and address of the named insured on the policy
- Issue date of the COI
- Name of the insurance carrier providing coverage
- A list of policy numbers for the various policies shown on the certificate
- Limits of liability that might be present, but for workers' compensation there are no true limits
- Effective dates and expiration dates for the various policies on the certificate
- Information as to whether the owner, partners, LLC members or corporate officers are included or excluded from workers' compensation coverage
- Certificate holder information, name and address

- Policy cancellation wording, conditions of notification
- The insurance producers' signature
- Additional insured notification for the certificate holder

You should ask your suppliers and anyone working on your property to regularly show proof of insurance. The certificate is written proof that workers' compensation and other coverage is being provided for a specified term.

It is important to evaluate company risks and mitigate these through contracts and COIs. Safety professionals should get actively engaged to protect property and employee exposures pointed out here.

In this ever-changing modern world, company strategies change constantly in order to respond to changing environments and regulations. Safety goals and strategies need to change and align with a company's goals and strategies for the company's continued success. For example, the terrorist attacks of 9/11 required companies to focus on improving the protection of their operations and facilities. This required safety to focus on security and training associates in spotting potential terrorist attacks both within and outside the plant operation.

World-class safety companies align their safety departments with various operations and sales departments and provide an adequate financial budget to secure meaningful results. Similarly, such companies recognize production achievements and sales goals along with safety goals. They are unwilling to recognize plant achievement awards unless there is a concurrent safety achievement as well.

The most important purpose of a business is to make profits for its owners and shareholders so that they can continue to operate and gainfully employ people. The most important purpose of employee safety initiatives is to make sure all employees go home safely to their families. It is important that safety is run as a business so that we can integrate the roles of business and safety. All safety meetings should be run like business meetings with a purpose and an agenda. The goal for safety professionals should be not only to run safety meetings like business meetings, but also to make sure senior executives realize and understand that safety meetings will be run as business meetings with a purpose and an agenda.

Every year at FDC I invited safety team leaders from all of the plants for a day and half of safety meeting at our corporate office in Elk Grove Village. One year, I called the CEO asked him if he wanted to address the safety champions from each of his plants. He said he was busy and would not be able to attend. I then put together a tight schedule for the day and half meeting. I had invited a guest speaker for the meeting to discuss behavior-based safety and the importance of making job observations so that safety and human resources leaders could take actions to correct hazardous acts and conditions

Unexpectedly, the CEO decided to pop into the morning meeting. For more than half an hour he talked about his business strategy. Finally, I had had enough! I got up and said, "I am sorry, but we have a tight schedule." Flushed and red-faced, he left and I continued the meeting. At lunchtime, I mentioned to my direct report, VP of operations, that this was a business meeting and he was invited to attend, but he declined.

The following agenda was distributed to the CEO prior to the meeting:

Agenda for Safety Summit
September 13th, 2009

- Introduction – Mike Saujani
- Safety Leaders in Total Safety Culture – Mike Saujani
- Break
- Train the Training – Training Director

- Ergonomics – Material Handling Assessment – Mike Saujani
- Lunch
- Safety Team Leader's Presentations
- Insurance Company Presentations
- Breakout Session – Important Safety Issues

Agenda for Safety Summit
September 22nd, 2010

- Introduction – CEO
- Safety – It is not just a job – Skipper Kendrick
- Lunch
- Panel Discussion
- FDC Company 2010 Accomplishments – Celebrate – Mike Saujani
- Safety Team Leader's Presentations
- Questions and future goals

He came back a five o'clock and saw that the participants were still working diligently to complete their assignments. From then on, the CEO came every year to open the meeting and encourage us in our initiatives. He became a champion for safety in the corporate boardroom. The following year, I received the biggest bonus of my life—more than my annual income!

The lesson here for safety professionals is to make sure you run your safety meetings as business meetings with a clear purpose and an agenda. Keep your CEO in the loop, but maintain control of your meetings. From that year on, our CEO always started the safety summit meeting encouraging participants to be proactive in promoting safety.

Data-based Decision Making and System-based Root Cause

Jeff Bezos turned to data-driven management very early. World-class companies manage based on the data they create. Bezos wanted his grandmother to stop smoking, he recalled in a 2010 graduation speech at Princeton. He didn't beg or appeal to sentiment. He just did the math, calculating that every puff cost her a few minutes. "You've taken nine years off your life!" he told her. She burst into tears and changed her ways.

It is important to gather data so that the safety system can be reviewed and analyzed using various analytic techniques. This will ultimately lead to making intelligent safety decisions, which might involve changing safety goals or strategies or realigning safety functions. Some of the data that world-class safety companies gather include:

- Incident rates, DART rates, loss time incident rates
- NCCI MOD
- Trends analysis for incident, DART, and loss time incident rates
- Loss analysis trends

- Gap analysis for established safety goals
- Safety awareness scores
- Hazard surveys
- Leading and lagging indicators

Accident Data

NCCI MOD

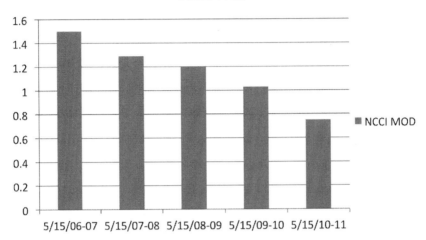

2001 - PRESENT LOST WORKDAY RATE

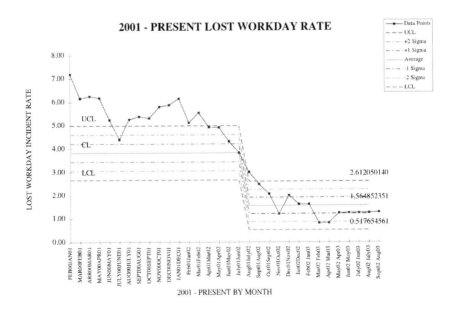

The National Council of Compensation Insurance (NCCI) mode may show a company's overall loss experience in a three-year period. For example, the MOD for the year 5/15/10-11 shows the average for the previous three years and the MOD for the year 5/15/09-10 shows the average for the prior three years, and so on. In the NCCI MOD chart above, the data reflects how well the company is doing in managing incidents and the cost of injuries.

The OSHA incident rate, Days Away/Restricted or Transfer (DART) rate, and the Loss Work Day Incident (LWDI) rate when an employee is unable to work a full assigned work shift reflect how many incidents have occurred in that category in a year on average per every hundred employees. It reflects a company's past performance and helps the company evaluate how well its safety system is performing compared to prior years and its peers.

It is one thing for executive management to say the company has world-class safety culture and another for the associates to say it. Hence, the importance of a perception survey to gauge what the perception is at the plant-floor level. Form a committee to select the right questions and oversee completion of the survey. The result and responses can be analyzed to determine where and how safety initiatives can be improved. A high rating at one plant and a low rating at another should reflect that the company safety goals and strategies are not well communicated throughout the organization.

NCCI MOD and OSHA incident rates are all lagging indicators of what went wrong, and they allow us to make the necessary adjustments to prevent similar incidents from occurring in the future. In order to proactively prevent accidents and injuries from occurring, it is critical for safety professionals to identify and monitor whether "Leading Indicators," as commonly used by safety professionals, have been followed, such as:

- Behavior-based safety (BBS) observations completed
- Safety training completed for managers and associates
- Safety audits completed and hazards identified

- Hazards identified and corrected
- Department safety meetings completed
- Coaching and counseling sessions held

Managing the leading indicators listed above will help your company proactively manage incidences and the cost of injuries and develop a culture of proactively managing safety.

Safety metrics: All businesses use metrics to predict and measure future performance. One of the metrics a commercial printing company uses is On Time in Full (OTIF). OTIF means the customer's printing labels should be completed on time and as ordered. It measures whether or not the production floor is able to accomplish this task. The higher the OTIF percentage, the better able the production floor is to complete customer order properly. If the OTIF is lower than expected, then an investigation is done to determine what went wrong and what should be done in the future to make sure the customer order is completed on time and in full. By the same token, we in safety have a safety operation matrix to measure safety performance. We measure the monthly and year-to-date recordable incidents, IRs, DART rates, and workers' compensation incurred cost at each plant as reflected in the metrics on the facing page.

Safety Operation Matrix December Summary

	Recordable Incidents			Incident Rates (IR)			Lost Time Incidents			Lost Time Incidence Rates (LWDII)			WC Incurred Cost	
	Current Month	Year To Date	YTD Last Year	Current Month	Year To Date	Last Year	Current Month	Year To Date	YTD Last Year	Current Month	Year To Date	Last Year	Year To Date	Last Year
Plant 1	1	12	18	6.5	5.3	7.8	0	1	0	0	0.4	0	$38,093	$136,438
Plant 2	3	9	7	25.3	6.1	4.7	0	2	2	0	1.3	1.3	$15,600	$92,903
Plant 3	0	3	4	0	2.8	3.4	0	0	0	0	0	0	$1,093	$42,677
Plant 4	0	13	12	0	12.8	13.9	0	0	1	0	0	1.2	$48,129	$22,314
Plant 5	0	0	0	0	0	0	0	0	0	0	0	0	$0	$0
Plant 6	0	0	5	0	0	9.5	0	0	2	0	0	3.6	$0	$3,670
Plant 7	1	10	15	7.3	5.5	7.7	1	3	4	7.3	1.6	2.1	$9,013	$53,212
Plant 8	3	6	1	27.7	5	0.8	1	2	0	9.2	1.7	0	$7,914	$9,624
Plant 9	0	1	0	0	0.5	0	0	0	0	0	0	0		
Plant 10	0	4	10	0	3.9	7.3	0	3	4	0	2.9	2.9		
Total	8	58	72	8.6	4.5	5.3	2	11	13	2.1	0.9	0.74	$119,842	$360,838

Note: The incidents and incident rates (IR, LWDII) are based on the calendar year while the workers compensation cost is based on Policy Year beginning June 1st.

One of the most important responsibilities of the safety team leaders is to complete the trending report on employee injuries for each of the plants they represent. They typically complete a trend report at a certain frequency using IR and LWDII rates. You can use the Process Control Chart Tool Kit (PCCTK) developed by Soft-Ware Tools or other similar programs, such as the QI Macros for Excel to monitor the safety process. The PCCTK and QI Macros are typically used by quality control and production departments to track various operations and processes throughout the organization, so it should not be difficult for safety professionals to use them. Using the PCCTK program, I used to monitor the safety process using the XbarR average and range chart and the XmR individual and moving range charts to evaluate the safety process's stability.[10] The XbarR chart provides the average rates and the upper and lower control limits, while the XmR chart provides process stability information. These charts help analyze whether there are any "Special Causes" influencing the safety process so that you can jump on the situation as soon as possible.[11] There are eight tests, four of which are listed below.[12] These are the most important for analyzing whether your company's safety process is stable and predictable. If the data on the chart violates any of the following four basic tests, the process is unstable and should be evaluated for special causes:

a.   1 point above or below the UCL or LCL
b.   8 consecutive points grouped above or below the average
c.   6 consecutive points ascending or descending
d.   14 consecutive points alternating up and down

Shown below is a typical LWDI chart from a plant. The interest-

---

10   J. Heitzberg, "Process Control Chart Tool Kit," *IBM Reference Manual for Windows*, version 6.0.2. (1985-1997): QI Macros, 14-17.

11   Peter R. Scholtes and Russell L Ackoff, *The Leader's Handbook: Making Things Happen, Getting Things Done* (New York: McGraw-Hill, 1997): 28.

12   J. Heitzberg: 147.

ing thing I discovered was that whatever gets measured actually gets done! This chart shows that the company's focus on safety paid off.

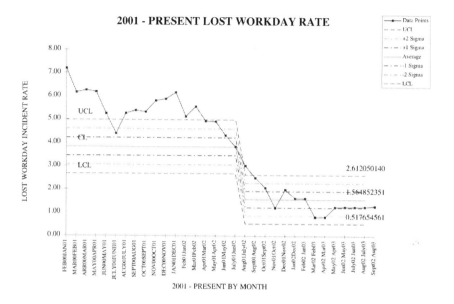

2001 - PRESENT LOST WORKDAY RATE

There are five major types of rates typically used when referring to employee injuries. The most common rate used is the recordable incident rate. This is commonly called the total case incident rate or just the incident rate. The Lost Time Case (LTC) rate is the second most commonly used. The LWDI and severity rate are primarily used for bigger companies that have a larger number of lost time cases. The DART is a mathematical calculation that describes the number of recordable injuries and illnesses per every one hundred full-time employees that result in days away from work, restricted work activity, and/or job transfer that a company has experienced in any given time frame. In a world-class company atmosphere, safety is an integral part of the operation and it swims or drowns with the operation. We started measuring the cost of employee injuries as well as IR and LWDI rates. This in turn helped us put in place other processes such as effective return to work and active claims management that helped us take care of our injured employees a little better. The operation matrix became a scorecard for safety and how well safety activities were managed.

Another interesting measuring tool for the system in a world-class safety culture is the balancing wheel. The wheel is divided into five segments, each based on the weight assigned to the Five Pillars of Safety. A random sample of people is taken to determine their perception of where the company is as regards each pillar of safety on a score of one to ten. Draw a cross across each pillar based on the score it received and then connect the lines between the crosses. The resulting figure tells you the state of the safety culture in your organization. It also helps you gauge how big the gap is between each pillar of safety. The goal is to have a score of ten for each pillar of safety so the wheel can move smoothly down the hill.

Balance Wheel Measuring Tool

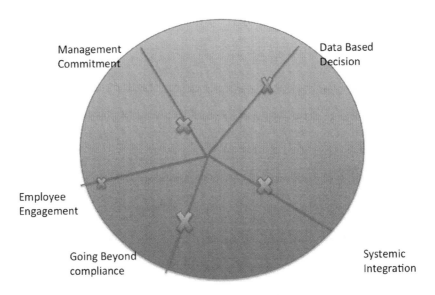

It is critical to measure safety data. That is a given. However, how we measure it and how we use the data is critical in our success as safety professionals. There are several safety software programs available in the market place to collect safety data from a company's process. These include Safety Management Software by Intelex, (www.intelex.com), Rivo Software (www.rivosoftware.com), Field iD by Master Lock (www.capterra.com), and others that allow the monitoring of safety incident reporting, OSHA Injury reporting, OHSAS 18001 compliance, audit inspections, training and competence, and safety data sheets. Several large companies with plants all over the world use these programs. Some of the files discussed here should help small to medium size companies in monitoring the safety process. At the same time, these programs do not monitor lagging and leading indicators. We will discuss how these can be monitored as well. Some of the elements of the safety process data I plan to discuss are:

- Operation matrix
- Training matrix
- Incident rates
- Lagging and leading indicators
- Loss analysis trends
- Hazard surveys
- Employee safety awareness
- Behavior based safety observations

Safety Operation Matrix: this is the matrix we used to gauge various items that reflect the safety pulse per plant and company wide. We tracked real-time incidences, monthly running trend of incidences and incidence rates compared to the company average. We also used the same matrix to gauge the cost of injuries per facility, compare the rates at different facilities with each other, and trend these rates by month according to historical rates and the NCCI industry average. The matrix provides the overview of the injuries by cases and costs per plant and company wide.

Training Matrix: this helps safety professionals plan and execute the training of associates for the year. At the beginning of the year, you list and identify the training you plan to complete and when.

This is listed on the X-axis of the matrix, while the names of the associates to be trained are listed on the Y-axis. When the training with the quiz is completed, the safety professional records it with a cross indicating successful completion as shown in the table below.

Safety Training Matrix

January 1, 2006								
Name	Hire Date	CPR	First Aid	Workplace Violence	Fire Safety & Evacuation	Bloodborne Pathogens	HazCom	Lock Out – Tag Out

When making a decision regarding the safety-related training of associates and managers, you can use the following matrix.

Type of Training	Associates Involved	Recommended Frequency	Comments
10-hour OSHA Certification	Managers	Once every Three years	Training presented by Corporate Safety and/or Insurance Consultant
Fork Lift Safety	Authorized Fork Lift Operators	Annual	
Hazard Communication	Affected Associates and all Managers	Annual	Training presented by managers & Safety Team Leaders
Lock Out Tag Out	Per OSHA Standard	Annual	Managers & Associates that do any service and repair
Ergonomics & Material Handling	Managers & Affected associates	Every Other year	Training presented by Insurance Consultant
Workplace Violence	All Associates	Every Other year	Safety Team Leaders and HR Managers to present the training

Type of Training	Associates Involved	Recommended Frequency	Comments
Fire Safety & Evacuation	All Associates	Annual	
Hazardous Waste Operation	Selected associates	Every Other year	Training presented by technical consultant from the waste hauler
Electrical Safety	Selected Associates	Every other year	Managers & Maintenance Staff
New Employee Safety Orientation	New employees	With in three months of hire	Can be completed by HR informally at the time of hire
First Aid/CPR/AED	Selected associates	Every Other year	Within 911 area code, at the discretion of GM
Blood Borne Pathogens	All Associates	Annual	Training by Department managers
PPE	All Associates	At Hire & continuous	Training by department managers
Accident Investigation and Reporting	All Managers	Annual	Corporate Safety or Insurance Consultant
Early Return To Work	All Managers	Every Three years	Corporate Safety or Insurance Consultant

Use this only as a guide. The goal should be to go beyond what is required by OSHA standards. Managers and safety team leaders should also review the OSHA standards to be certain their training guidelines meet them and other standards as well.

Employee Awareness Surveys: it is important that we survey our associates and evaluate how engaged they are in the safety process. I suggest doing a survey of how engaged the associates are in safety using a checklist at least once a year. The attached checklist is an example that can be used to measure employee perceptions. I have provided completed sample forms to help you properly and effectively use this method of measuring the employee pulse. One sample is the form completed by employees, the other is a summary of all forms completed for that plant, and then a summary of the forms completed by all plants, which in this case included six plants.

SAFETY SURVEY

We are undertaking a number of initiatives aimed at raising safety standards. The company has decided to focus on employee attitudes and perceptions as one of these initiatives.

To help with this task, we would like you to complete the following questionnaire—confidentiality is assured. We ask only for basic job information in order to help interpret the results.

The questionnaire is relatively simple to complete and asks about your attitudes on safety issues; as well as any suggestions you might have to improve things.

Please answer all of the questions.

Thanks for your assistance.

It is important for you to be completely honest about your feelings. All responses will be treated in strict confidence and there is no requirement to put your name on the questionnaire. The responses will be recorded in confidence by the corporate safety director.

It should take 10 to 15 minutes to complete this questionnaire.

We would like you enter your plant and department to assist us with the interpretation of the results.

Thank you for your cooperation.

FDC SAFETY SURVEY

Plant: _____

Department: _____

You will be presented with a series of statements about health and safety. Please indicate your response by marking the appropriate answer for each question:

1 Strongly disagree	2 Disagree	3 Neither agree nor disagree	4 Agree	5 Strongly agree			
Questions				**1**	**2**	**3**	**4**	**5**
1. Management operates an open door policy on safety issues.								
2. Safety is a priority in my mind when completing a job.								
3. Co-workers often give tips to each other on how to work safely.								
4. Safety rules and procedures are carefully followed.								
5. Management clearly considers the safety of employees of great importance.								
6. I am given enough time to get the job done safely.								
7. I am involved in informing management of important safety issues.								
8. Management acts properly when a safety concern is raised.								
9. There is good communication here about safety issues which affect me.								
10. I understand the safety rules for my job.								
11. It is important to me that there is a continuing emphasis on safety.								
12 This is a safer place to work than other companies I have worked for.								
13. I am strongly encouraged to report unsafe conditions.								
14. Personally I feel that safety issues are not the most important aspect of my job.								
15. In my workplace the chances of being involved in an accident are quite large.								
16. I do not receive praise for working safely.								
17. I can influence health and safety performance here.								
18. When people ignore safety procedures here, I feel it is none of my business.								
19. I am clear about what my responsibilities are for health and safety.								

Questions	1	2	3	4	5
20. A safe place to work has a lot of personal meaning to me.					
21. There are always enough people available to get the job done safely.					
22. In my workplace managers/supervisors show interest in my safety.					
23. Management considers safety to be equally as important as production.					
24. Managers and supervisors express concern if safety procedures are not adhered to.					
25. I cannot always get the equipment I need to do the job safely.					

Do you have any other comments about health and safety in your workplace?

COMMENTS FROM ASSOCIATES

1. "This Company has shown concern for my safety and wellbeing. It is very nice and important that safety issues are looked into."
2. "The production worker should change their habits of standing empty pallets on their side as these could fall on someone."
3. "I feel everyone I've worked with has worked in a responsible manner—always with safety in mind."
4. "Management has provided us with proper equipment for our welfare and safety."
5. "I think this is a very safe place to work."
6. "I think the company does a great job of ensuring employee safety."
7. One associate said, "I have no problem working here because it is really safe here."
8. "The only concern I have is proper use of chemicals and how it would affect me in the future."
9. "Here, safety is always 1st."
10. "I feel the company is giving 100 percent in this (safety) area."
11. "I'm concerned about the current ventilation system and lack of humidity in the winter."

12. Another associate said management thinks safety is very important.
13. One associate said that there is a strong emphasis on safety; however, quality job done quickly is priority number one.
14. "It is time to set safety guidelines for each department and each piece of equipment."
15. "We watch out for our associates."
16. "The associates should get the pat on the back and not the managers for our safety record."
17. "Management here always puts safety first."
18. "Everyone works together as a team to keep this place safe."
19. "Safety is number one here."
20. "There is good and regular emphasis on safety here."
21. "Make safety a fun thing. Keep using the games and observation as safety prizes."
22. "We have a real good safety program."
23. "Signs and mirrors have helped prevent accidents."
24. "Working safely leads to great rewards at work and also in our everyday life."
25. "A safe team that works together wins together."
26. "Having a luncheon to celebrate safety in the workplace is very positive feedback."
27. "A common sense safety feature is wearing safety shoes, glasses and other personal protective equipment."

In order to make sure the survey offers an accurate reflection of what the employees feel, it is important to use random sampling. I would recommend asking HR to provide a list of all employees and then select 12-15 percent at random to take the survey. You may even want to use random numbers in selecting the list of employees. The associates completing the survey then mail their responses directly to the corporate office where they are tabulated and scored on a scale of one to five. The following were the results of a survey that I completed at FDC:

Plant 6 – 4.3
Plant 1 – 4.1
Plant 4 – 4.0
Plant 2 – 3.7
Plant 5 – 3.7
Plant 3 – 3.6

The scores and comments were reviewed with each manager at the plant for improving associate engagement.

World-class companies collect accident data for various reasons. The most important reason is to identify ways to prevent accidents from occurring in the future. Then those who collect the data determine whose fault it was so that the cost of the accident can be properly allocated to the source. Others collect data so that an analysis can be conducted and an improvement to the process can be recommended. We also collect data in order to comply with government and OSHA reporting requirements. Generally, the acceptable method of collecting employee injury data is using the state first report of injuries form.

ILLINOIS FORM 45: EMPLOYER'S FIRST REPORT OF INJURY

Please type or print.

Employer's FEIN	Date of report	Case or File #	Is this a lost workday case? Yes No
Employer's name		Doing business as	
Employer's mailing address			Employer's email address
Nature of business or service			SIC code
Name of workers' compensation carrier/admin.		Policy/Contract #	Self-insured? Yes No
Employee's full name			Birthdate
Employee's mailing address			Employee's e-mail address

Gender Male Female	Marital status Married Single	# Dependents	Employee's average weekly wage
Job title or occupation			Date hired
Time employee began work	Date and time of accident		Last day employee worked

If the employee died as a result of the accident, give the date of death.	Did the accident occur on the employer's premises? Yes No
Address of accident	

What was the employee doing when the accident occurred?

How did the accident occur?

What was the injury or illness? List the part of body affected and explain how it was affected.

What object or substance, if any, directly harmed the employee?

Name and address of physician/health care professional

If treatment was given away from the worksite, list the name and address of the place it was given.

Was the employee treated in an emergency room? Yes No	Was the employee hospitalized overnight as an inpatient? Yes No	
Report prepared by	Signature	Title and telephone # Email address

Please send this form to: ILLINOIS WORKERS' COMPENSATION COMMISSION 4500 S. SIXTH ST. FRONTAGE RD SPRINGFIELD, IL 62703
By law, employers must keep accurate records of all work-related injuries and illness (except for certain minor injuries). Employers shall report to the Commission all injuries resulting in the loss of more than three scheduled workdays. Filing this form does not affect liability under the Workers' Compensation Act and is not incriminatory in any way. This information is confidential. IC   '12

World-class companies go beyond what is required by state or OSHA; they require their managers and supervisors to complete an accident (incident) investigation report after completion of a thorough investigation of the accident. Various types of forms have been used to gather accurate information leading to the accident. For example, a supervisor's accident investigation form could look like the form shared here.

Managers: Please fill out this form after the associate has completed the reverse side.

1. Employment Category: Full-time _____ Part-time ____ Temporary ____ Seasonal ____ Non-employee _____	2. Length of Employment: Less than 1 month _____ 1 to 5 months _____ 6 months to 5 yrs _____ More than 5 yrs _____	3. Time in Occupation at time of incident: Less than 1 month _____ 1 to 5 months _____ 6 months to 5 yrs _____ More than 5 yrs _____

4. Type of Operation: Production _____ Cleaning _____ Maintenance _____ Unjamming _____ Set up _____ Other _____	5. Was this operation a regular part of the employee's job? Y____ N____ 6. Time Associate reported to work on date of incident: _____

7. Type of Injury: No Injury__ First Aid__ Medical Treatment__ Lost Days__ Restricted Days__ Fatality__

8. Equipment have lockable disconnect? Y__ N__	9. Name and Title of person who assigned job:

10. How long since a Supervisor was in area and who? _____

11. Accident Sequence: (Describe in reverse order of occurrence events preceding the injury and accident. Starting with the injury and moving backward in time, reconstruct the sequence of events that led to the injury.)
A. Injury Event _____
B. Accident Event _____
C. Preceding Event #1 _____
D. Preceding Events #2,3 etc. _____
_____

12. What PPE was required?	Provided?	Trained?	Worn?	Enforced?
	Yes / No	Yes / No	Yes / No	Yes / No
	Yes / No	Yes / No	Yes / No	Yes / No
	Yes / No	Yes / No	Yes / No	Yes / No

13. Was equipment adequately guarded? Y__ N__
If 'No', what was not correct? _____

14. Had Associate received training prior to job assignment? Y__ N__	Was it adequate? Y__ N__
Who performed the training? _____	What was the duration? _____

15. Was lockout necessary? Y__ N__	If yes, was Associate trained? Y__ N__	Adequately? Y__ N__
Was a padlock provided? Y__ N__	Was written procedure required? Y__ N__	Provided? Y__ N__

16. Conclusion/Root Cause of incident: _____

17. Corrective Actions: Those that have been, or will be taken to prevent recurrence of the injury and/or accident. _____
_____
_____

18. Supervisor:	Date:
Department Mgr:	Date:
Safety Officer:	Date:
Office use only: Policy # 273153 Date filed _____ with _____ Reference # _____	

It is important that all managers receive guidance and training in completing these reports for a successful outcome. The following flow chart can be used to provide guidance when investigating an accident.

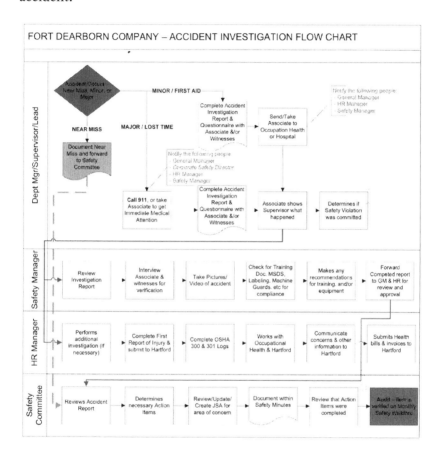

The investigation report should cover the following topics:

PURPOSE:

To complete a thorough investigation of an accident.

SCOPE:

This procedure involves investigating accidents so as to obtain factual information in preventing future accidents.

RESPONSIBILITIES:

The department manager, HR manager, and the safety team leader will investigate all accidents. In addition, the general manager will investigate all lost time accidents or property loss.

ACCIDENT INVESTIGATION PROCEDURE:

It is important to prevent accidents for the safety and welfare of our associates. Therefore, when an accident does occur, it is very important that we learn from our mistakes and identify the causes of the accident so that these can be controlled. The accident investigation procedure should include:

a. Completion of the accident investigation form by the injured associate, department manager, safety team leader, and general manager for the lost time accident or property loss.
b. Interview of the injured associate as well as associates who witnessed the accident, preferably at the scene of the accident.
c. Taking pictures or video of the scene of the accident.
d. Treating the injured associate with care and concern.
e. Determining the root cause, primary causes, and the basic causes of the accident.
f. Recommending the corrective process improvements necessary to prevent similar accidents in future.
g. Communicating unrecognized or uncontrolled hazards to the general manager for action.

Accident data gathered in this way can then be used to analyze accidents and near misses to determine the cause or causes of an accident or series of accidents so as to prevent further incidents of a similar kind from occurring. The accident data can be grouped in a variety of ways to understand causes of similar accidents so that corrective changes can be implemented to various processes causing these accidents. These can be grouped by type of body parts injured, type of injuries, condition leading to the injuries, etc. An example of the accident analysis can be seen below.

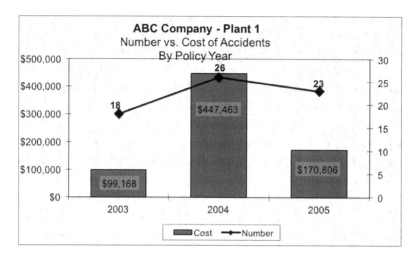

One might conclude from this analysis that the number of injuries has increased within the last four years; however, the cost of injuries has gone down. These results could mean that the safety professional needs to concentrate on the frequency of injuries and find ways to minimize the frequency. Here is another example analysis.

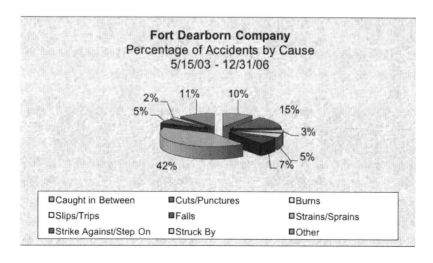

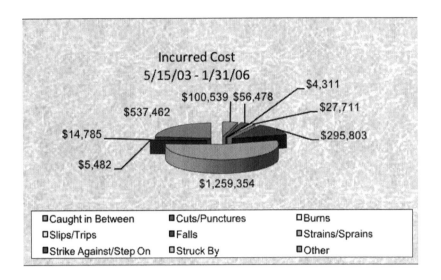

Incurred Cost
5/15/03 - 1/31/06

$4,311

$100,539 $56,478

$537,462

$27,711

$14,785

$295,803

$5,482

$1,259,354

▣ Caught in Between	▪ Cuts/Punctures	▢ Burns
▢ Slips/Trips	▪ Falls	▣ Strains/Sprains
▪ Strike Against/Step On	▢ Struck By	▣ Other

Accident Statistics
May 15, 2008 – March 15, 2011

Type of Injuries	# of Injuries	% of Total	Cost of Injuries	% of Total
Strains and Sprains	103	45%	$669,258	69%
Hernia	2	1% <	$29,709	3%
Multiple Injuries	34	15%	$195,379	20%
Cuts and Laceration	60	26%	$45,623	5%
Miscellaneous	29	13%	$31,633	3%
TOTAL	228	99%	$971,602	100%

These two analyses cover a two-and-half-year period from May 14, 2003 to January 31, 2006, and show that the ergonomic injury cost was $1,259,354 out of about $2.5 million in total injuries or more than 50 percent of the total cost for this two-and-half-year period. The company's safety professional decided to improve the ergonomic program as a result of this analysis. A few years after implementing an ergonomic program, while analyzing the cost on March 15, 2011, the cost of injuries was found to have been reduced to $669,258, as shown in the second chart, a reduction of $590,096. That is a significant reduction in cost and improvement of the ergonomic hazards to employees. One of the data sets gathered in enhancing ergonomic process improvement was enhanced accident data for ergonomic-related injuries.

Ergonomics is a science that seeks to adapt tasks and tools to fit the person. It is a value added process that identifies and removes barriers between people and production and improves quality and safety. It is the practice of designing jobs and workplaces to match the capabilities and limitations of the human body. Our goal is to improve quality and productivity, employee health and safety, comply with regulatory requirements, and reduce the cost of injuries. The ergonomic program that one company put in place consisted of the following:

- Management commitment/roles and responsibilities
- Ergonomic training
    - Understanding ergonomics and risk factors
    - How to complete task analysis
    - Communication of ergonomic related information
- Ergonomic task analysis
- Job improvements and brainstorming
- Ergonomic injury investigation
- Ongoing program support/safety audit
- Record keeping

The program reduced ergonomic-related injuries significantly as shown above.

NCCI, MOD, Loss Analysis, and OSHA incident rates are all lagging indicators that help establish what went wrong and allow us to make necessary adjustments to prevent similar incidents from occurring in the future. In order to proactively prevent accidents and injuries, it is critical for safety professionals to identify and monitor "Leading Indicators" such as:

- Behavior-based safety observations completed
- Safety training completed for managers and associates
- Safety audits completed and hazards identified
- Hazards identified and corrected
- Department safety meetings completed
- Coaching and counseling sessions held

Managing these necessary actions will help your company avoid accidents and the cost of injuries and develop a culture that proactively manages safety. Management should regularly collect information on lagging indicators, records that show what went wrong and how a procedure can improve moving forward. The lagging and leading indicators can be measured by several simple systems or charts and help us learn from our mistakes. However, for lagging indicators this learning happens after the fact, after an incident has occurred. There is a positive correlation between incidents and leading indicators, something you do before an incident occurs such as employee training. Leading indicators are activities employees do every day to prevent injuries or accidents, which includes safety committee meetings, safety training, hazard evaluation, safety observations and conversations, and other similar activities.

There is a relationship between these leading indicators and accidents. If you do not train your employees or identify and correct hazards on a regular basis, it is highly likely that employees will get hurt. In fact, some of the OSHA standards require companies to put in place leading activities to prevent accidents, such as regular crane inspections and employee safety training on lift-truck operation. World-class companies are actively involved in monitoring leading indicators. One such company uses a simple chart to monitor and ensure that these activities are being completed throughout the organization.

Lagging and Leading Indicators
Month of March 2012

	Incidents	Lost Time	Training	Observations Completed	Inspections	Safety Meetings	Dept. Meetings	Plant-wide Safety	Safety Recognition
Plant 1	3	1	2	5	2	1	4	1	2
Plant 2	1	1	1	7	1	1	3	1	
Plant 3	4		1	4	2	1	3	1	1
Plant 4	0		2	8	3	1	2		1
Plant 5	3		2	3	2		4	1	1
Total	11	2	8	27	10	4	16	4	5

Hazard evaluation and the correction of identified hazards is critical for sustaining safety. Self-inspections, inspections completed by employees and immediate supervisors, prevent situations that have the potential to injure people or damage property. There are two types of inspections: informal and formal. At present, we will talk about the informal inspection only. In an informal inspection, a manager or an associate constantly reviews his operation for conditions or practices that might cause property damage or bodily injury. After an informal inspection, the department manager should state in the department inspection form whether any physical hazard was noted during the inspection, and fill out a work order for correcting the hazard if necessary or correct the hazard immediately and note that in the form. If the manager notices any unsafe work practices, then he should review the unsafe practice with the associate in question and offer him coaching or counseling. If an associate notices any unsafe conditions or actions, he should bring it to the attention of his supervisor. A sample forms that can be used for this purpose is included here:

Plant 1 – Safety Inspection Checklist

PROCESS SAFETY:

1. Nothing stored in or on the white 18" from wall around the perimeter of the production area.
2. Process lines / pipes labeled.
3. Eye wash stations accessible and operable.
4. First-aid stations accessible and operable.
5. Fire extinguishers fully charged, accessible and not obstructed from view.
6. Fire bells checked monthly.
7. Chemical-soaked rags placed in approved safety cans/receptacles.
8. No obstacles blocking walkways/aisles – aisles clear and free of tripping hazards.
9. No bare wire exposed on electrical cords, fixtures or outlets.
10. Nothing within 18" of ceiling (sprinkler system parameter).
11. Ceiling tiles all in place / none missing.
12. Electrical cords properly covered in walkways.
13. All electrical cabinets are closed.
14. No food consumed (or gum chewing) in the production area.
15. No open drinking vessels in the production area.
16. Work areas clean and free of chemical spills.
17. Pallets stacked safely with no materials falling over the edges.
18. No nails protruding from wooden pallets.
19. Racks and/or drums used for storage of flammable liquids are properly grounded.
20. Metal cases of electrical equipment are properly grounded.
21. Liquid spills are cleaned up immediately.
22. Exit signs clearly visible and well illuminated.
23. Exit and/or emergency doors are free of obstruction, accessible and unlocked for egression.

24. Fire extinguishers are checked visually monthly and recharged annually.
25. No loose razor blades lying free.
26. Receptacles available for disposal of used razor blades.
27. MSDS sheets available and accessible to laboratory and plant personnel.
28. Forklifts operated in a safe manner.

PERSONAL PROTECTION:

1. No loose jewelry worn by press/production personnel while operating moving-part machinery.
2. Safety glasses/goggles worn in production area when transferring/pouring liquid chemicals.
3. Protective gloves worn by production and lab personnel when working with hazardous inks/chemicals.
4. Press operators / production personnel utilizing proper lifting technique. (Keep back straight, lift w/ leg muscles)
5. No hanging sleeves, shirttails, loose-fitting clothing, or (unsecured) long hair in the press area.
6. Safety shoes worn by all production personnel (Digital/Flexible Packaging).
7. Availability of gloves for production personnel.
8. Operators/pressman utilize cutting blades in such a fashion as to cut *away* from themselves when using razors.
9. Emergency evacuation procedures in place.

STORAGE AND LABELLING OF CHEMICALS:

1. There are NO unlabeled containers in the lab or production area.
2. HMIS labels w/ corresponding hazard ratings are on all chemicals in both the lab and production areas.
3. THF drums grounded and maintained in secondary containment.

EMPLOYEE TRAINING:

1. Employees are trained on how to read an MSDS and know where MSDS are located, as well as labeling requirements and hazards associated with hazardous chemicals (OSHA's Hazard Communication Standard).

2. Employees are trained on HMIS hazard rating system.

3. Individuals working with THF (tetrahydrofuran/tetrabright) have received additional training on the hazards and proper handling of THF.

4. Bloodborne pathogen training completed for all manufacturing managers and training documented.

5. Employees trained on emergency evacuation procedures.

Please add or subtract items that apply or do not apply to your operation. Informal inspections require using the checklist for inspection but not necessarily completing and storing the form. Just make sure the hazards noted are corrected immediately or a work order for their correction is developed.

The following Lift Truck Safety Inspection form can also be used for formal and informal inspection of the lift truck.

POWERED INDUSTRIAL TRUCK DAILY INSPECTION

An examination of powered industrial trucks shall be made daily, at the start of each operating shift. This examination will usually be made by the operator. Conditions adversely affecting the safety of the vehicle shall be reported and corrected prior to placement in service.

INSPECT	INSPECT
• Brakes (Service & Parking)	• Alarms, Directional controls
• Steering, gauges, tires, …	• Load Brackets Extension
• Warning Device (Horn & Light)	• Front End Attachment
• Overhead Protection	• Battery Cables
• Fire Extinguishers	• Data & Warning Decals
• Fuel/Hydraulic Leaks?	• Other

Date/Shift	OK?	Initials	Date/Shift	OK?	Initials
1.			16.		
2.			17.		
3.			18.		
4.			19.		
5.			20.		
6.			21.		
7.			22.		
8.			23.		
9.			24.		
10.			25.		
11.			26.		
12.			27.		
13.			28.		
14.			29.		
15.			30.		

Date	Deficiencies or Service Required	Date Corrected/Initials

Note: You may also use a slightly revised version of this form for inspecting other equipment and machinery.

Selected members on the safety committee may be asked to complete monthly safety inspections using the safety self-inspection checklist to identify unsafe actions and conditions so that plant engineering departments can take the appropriate corrective measures. Invite State OSHA inspectors to examine your operations for hazards so that you can take your company's safety to the next level. These inspectors will work with you to provide guidance and assistance in correcting any workplace hazards. However, you must address all of their recommendations, otherwise they are obligated to inform Federal OSHA, who can then come in and impose citations to mandate the appropriate actions on your part.

It is important that a safety director or an independent private consultant complete an in-depth safety audit of the company's operations in order to assist company executives in setting a long term safety goal. In the past, I have used the MOCK OSHA inspection checklist to complete a detailed safety audit and calculate the potential fines if it was indeed an OSHA audit. This helps senior management grasp the cost of non-compliance and motivates local management to correct the hazards identified. An inspection checklist should vary depending on the type of industry and the extent of the audit. An example of the checklist can be found below.

Date:		Name:		
No	Description of Standard	Standard	Violation	Penalties
	General Safety			
1	OSHA Workplace Safety	1910.1		
2	Emergecny Telephone numbers Posted	1910.38		
3	Summary of Illness And Injury Posted	1904.5		
4	OSHA 200 Log Current	1904.2a		
5	Written Safety and Health Program	1910.1		
6	Safety Responsibility Clearly Delegated	1910.1		
7	Regular Meeting of the Safety Committtee	1910.1		

No	Description of Standard	Standard	Violation	Penalties
Date:		Name:		
Electrical Hazards				
21	Ground Fault Protection	1910.304f		
22	Ground Fault Circuit Interrupters	1910.304.a		
23	UL Certified Portable Electrical Equipment Used	1910.305		
24	Flexible Cord of Good Quality	1910.305j		
25	Permanent Electric Wiring Used When Required	1910.304		
26	Controlled Access to Installation over 400 Volts	1910.308a		
27	Electrical Boxes and Panels Properly Covered	1910.305b		
28	No Splicing and Taping of Electrical Cords	1910.305		
29	Assured Eletrical Grounding Conductor Program	1910.304		
30	Improper Temeporary Electrical Connetions	1910.303		
Fork Lift Safety				
41	Only Trained and Authorized Operators	1910.178.1		
42	Lift Truck Operating Rules Posted	1910.178m		
43	Lift Trucks Adequately Maintained	1910.178q		
44	Battery Charging In Designated Area Only	1910.178g		
45	Overhead Guard Provided When Required	1910.178m		
46	Back Up Alarm Provided	1910.178m		
47	Safety Equipment Provided	1910.178.m		
48	Brake Pads in Good Condtion	1910.178		
49	Seat In Good Condition	1910.178		
50	Seat Belts in Good Condition	1910.178		
Flammable Liquids				
51	Adequaate Ventilation	1910.106e		
52	Grounding of Flammable Liquids	1910.106e		
53	Bonding of Flammable Liquids	1910.106e		
54	Any Sources of Ignition within 20 Feet	1910.106e		
55	Flammable Liquids Stored in UL containers			
56	Flammable Liquids in UL Listed Cabinets			
57	Flammable Liquids Storage Room	1910.106f		

No	Description of Standard	Standard	Violation	Penalties
Date:		Name:		
Printing Machines				
61	Point of Operation Guarded	1910.212a		
62	Machines Inspected and Checked Weekly for Safety			
63	Safe Operating Instructions Provided			
64	Transmission Drives Guarded	1910.212		
65	Employees Trained in Safety			
66	Electrical Interlock System in Working Order			
Hazard Communication				
71	Written Hazard Communication Program	1910.12		
72	Hazard Communication Training Provided	1910.1200h		
73	Material Safety Data Sheets Available	1910.1200g		
74	Material Safety Data Sheets Maintained	1910.1200g		
75	Hazard Warning System - MSIS - Implemented	1910.1200f		
76	Training Evaluation Completed?			
Lockout Tagout				
81	Lockout Tagout Procedures Developed	1910.147c		
82	Employees Trained in Lockout Tagout Procedures	1910.147c		
83	Mahinery De-energized while Cleaning	1910.147e		
Hearing Conservation				
91	Hearing Conservation Program Written	1910.95c		
92	Noise Survey Completed	1910.95h		
93	Approved Hearing Protection Available	1910.95h		
94	Annual Audiometric Testing Completed	1910.95g		
95	Hearing Protetors being Used.	1910.95h		
Storage And Handling				
101	Fall Hazard From Material Stored	1910.196		
102	Storage Area free of Trip and Fall Hazard	1910.176a		
103	Floor Opening Guarded	1910.23a		
104	Adequate Lighting	1910.178h		
105	Aisles Marked?	1910.22b		

World-class safety companies use Job Safety Analysis (JSA) to effectively reduce incidents, accidents, and injuries in the workplace. This procedure is an excellent tool to use during new employee orientations and training, and it can also be used to investigate near misses and accidents.

The JSA process starts by selecting the job or task that needs to be performed. Generally, tasks that have hazards or potential hazards or jobs that are uncommon or seldom performed are selected for JSA. The procedure includes multiple steps:

Basic Job Steps:

Break the job into a sequence of steps. Each of the steps should accompany some major task. That task will consist of a series of movements. Look at each series of movements within that basic task.

Potential Hazards:

To complete a JSA effectively, you must identify the hazards or potential hazards associated with each step. Every possible source of energy must be identified. It is very important to look at the entire environment to determine every conceivable hazard that might exist. Hazards contribute to accidents and injuries.

Recommended Safe Job Procedures:

Using the *sequence of basic job steps* and *potential hazards*, decide what actions are necessary to eliminate, control, or minimize hazards that could lead to accidents, injuries, damage to the environment or possible occupational illness. Each safe job procedure or action must correspond to the job steps and identified hazards.

As an example, please review the following:

Job Safety Analysis

Job Title:		Equipment:		Date Approved:
Director Assigned:			Revised Date:	
Persons Assigned				
Names: 1)	2)	3)		4)
Summary				
Task	Job Function		Tools & Materials	Hazard/Recommendations

Job Safety Analysis

Job Title: Packer		Equipment: Sortation System		Date Approved: 3/12/07
Director Assigned: Salim			Revised Date: 5/15/15	
Persons Assigned				
Names: 1) Anna	2) Linda	3) Laverne		4)
Summary				
Task	Job Function		Tools & Materials	Hazard/Recommendations
Assigning labels to lanes	Make ready		Computer, strip of labels and a marker	None
Moving skids of pre made boxes from warehouse to Finishing	Make ready		Pallet jack, cartons	Lifting, back pain, muscle strain and wrist pain/ apply ergonomics, wear back belt and wrist band.
Packing bundles in boxes	Packing off sortation		Cartons, paper bundles	Grabbing, bending and lifting/ apply ergonomics.
Packing bundles in boxes	Packing off cutting line		Cartons, paper bundles, tape gun and wooden pallets	Bending and lifting/ apply ergonomics.

World-class safety companies monitor the performance of their supervisors and managers constantly. It is a supervisor's respon sibility to be certain that employees are working safely; how well he or she performs this responsibility can and should be measured. The measurement of his or her performance can then be used to determine the level of financial benefit for the supervisor and manager. One such method is shown below.

Supervisors Safety Performance Review

	Rating Items	Adequate (2)	Average (3)	Excellent (5)
1	**Safety Meetings & Safety Training**			
	Conducts Monthly Dept Meetings			
	Attends Plant Safety Meetings			
	Attends Training Programs (Haz Com, etc…			
	New Employee Safety Orientation Provided			
	Completes monthly Tool Box Talks			
2	**Safety Audit**			
	Conducts Daily Informal Safety Audit			
	Are unsafe conditions corrected			
	Corrective work orders issued			
	Provides counseling and coaching			
	Observes associates work safely			
3	**Job Safety Analysis**			
	Completes Job Safety Analysis			
	Secures employee participation in JSA			
	Quality JSA prepared			
4	**Accident Investigation**			
	Completes Accident Investigation in 24 Hours			
	Thoroughly completes Accident Investigation Forms			
	Recommendations completed for Corrective Actions			
5	**Supervisors Safety Attitude**			
	Demonstrate sincere interest in safety			
Total Score				

Behavior-based Safety

For world-class safety companies, safety is a process; it does not have a beginning or an end. It is a process of continuous improvement. Behavior-based safety, also a process, reflects a proactive approach to safety and health management and takes into consideration four performance factors:

- The workplace
- The individuals who perform work
- Systems – outside and inside
- Human Relations

Part of the process involves identifying at-risk behavior through safety observations and physical audits and understanding the behavior that leads to accidents by conducting thorough investigations. Management must proactively prevent at-risk behavior through coaching and education. The goal is to maintain a positive, safe behaviors via training and motivational programs.

A typical form that can be used to perform behavior-based safety observations is available below.

Behavior-based Safety Observations

Name:	Date:	Department:
Subject: Safety		
Safety Observation – At Risk Behavior		
Proactive Safety Action – Coaching, Counseling, Training, etc… Date:		
Safety Committee Review: Date:		
General Manager Review:		

It takes a significant amount of preparation and implementation time for behavior-based safety is to succeed. You will have to spent the time and effort to make sure your program is designed to achieve the required results by training and encouraging your managers, supervisors, and employees who are involved in the process. Monitor your successes and failures so that appropriate follow-up actions can be taken to make it a successful program in engaging employees through coaching and counseling.

Other data not included in this chapter but that has been discussed in previous chapters in detail needs to be considered in evaluating and improving safety process so as to achieve world-class safety culture.

The data development and collection involving completion of Lock Out Tag Out procedure was discussed in detail in the Chapter 3 under Employee Involvement and Ownership.

The Safety Snapshot data showing company leadership involvement in getting safety results was discussed in Chapter 2 under Management Leadership and Commitment.

Minutes of the Safety Committee meetings, the data pertaining to the function and operation of the safety committee was discussed in detail in Chapter 3 under Employee Involvement and Ownership.

Please realize it is very important to gather meaningful data, analyze the data, and take actions that continuously improve the safety and manufacturing or service processes so as to achieve and sustain world-class safety culture.

Beyond Compliance

World-class companies go beyond compliance; that is their minimum requirement. Companies of this stature are proactive in identifying hazards and thoroughly investigate accidents and near misses so that the safety system can be constantly improved. They find ways to engage the entire organization, and they comply with OSHA or EPA regulations not only because it is required, but also because it is the right and profitable thing to do for their business. Many world-class companies invite their suppliers and vendors to provide feedback. A report from The Hartford, an insurance company, demonstrates its evaluation of safety culture at FDC:

> "Although there was a very strong commitment and formal program in place, the company was eager to reach out and learn in an effort to continually strengthen their program. Secondly, we quickly realized that this program had the full support of management. A management influence is present in all safety activities, and there is no doubt that safety is a key component of their business."

For a world-class safety company, safety is a team effort between senior executives, managers, associates, and safety professionals working together to create a safe and healthy environment. They run their business by continually improving safety processes and maintaining a high level of expectations for all stakeholders.

Active Involvement in Safety Associations

World-class safety organizations encourage safety professionals to actively engage in professional safety societies such as American Society of Safety Engineers (ASSE), which promotes professional development, expertise, and leadership by providing safety professionals with the knowledge they need to promote safety in their organizations. Safety societies not only assist in the development of expertise, but they also provide a platform for future learning and leadership. Associations allow professionals to network with peers who can be tapped later on for information and assistance.

For example, I was encouraged to join ASSE and actively participate in the association. Becoming an active member required a significant time commitment or about 8 percent of my available time. Initially I only attended meetings that could help me gain knowledge that I could apply to the work environment for the betterment of company employees. My peer group encouraged me to talk about my company's safety experience. I even invited members to tour our plant and discuss the safety initiatives underway there. As a result of the feedback I received from these members, I was able to improve our ergonomic safety program with significant financial savings for the company.

I later took on the responsibility of editing the ASSE newsletter, where I shared safety-related topics and news with chapter members. The confidence and knowledge I gained from this endeavor inspired me to develop a similar safety newsletter internally for our company employees. Eventually, I took responsibility for running the local ASSE chapter, which was recognized as a platinum chapter, the highest honor, in a letter from the president of ASSE:

Dear Mike,

It was a great pleasure to learn that the Northeastern Illinois Chapter will be recognized at the Chapter Recognition Luncheon as a Platinum Level Chapter. I know the bar is set high to achieve this status. The work you have put in is greatly appreciated by Society, and your members are the beneficiaries of your efforts. Hopefully, you feel rewarded for what you do, partly for the recognition, but mostly because you have made a tremendous impact on your chapter members and our profession. It is important that your achievements are recognized and let me assure you they have. Thank you.

Again, congratulations on your accomplishment, and thank you for being a role model for us all. I hope to see you in Atlanta!

Sincerely,

Michael Belcher, CSP
ASSE President

As a result of my work at ASSE and in my professional field, I received the Safety Professional of the Year Award (SPY) for the chapter and ASSE Region V. Not only was I recognized, but the company I worked for was also recognized as an outstanding safety company.

Safety Award Recognition

World-class safety companies realize the importance of encouraging safety professionals to actively engage in safety societies for the benefit of all.

Sustainability Practices

World-class safety companies also get actively involved in sustainability efforts—to sustain the environment for our children and future generations. They work diligently to reduce waste whenever possible.

At FDC we used the Four Pillars of Sustainability to promote sustainability practices. The four pillars are: materials, process, environment, and social responsibility. We will discuss only the environmental aspect of running a company's sustainability initiatives, which includes managing water waste, carbon footprint, VOC (volatile organic compounds), gas usage, and electrical usage.

In order to make any improvements, you need to be informed about wasteful practices and how best to conserve water, electricity, materials, and volatile organic compound emissions. Once you've observed a company's current practices, then you can determine areas where improvements are needed. The following list is an example of best practices.

CURRENT PRACTICES			Plant		
Area	Sustainability Practice	Comments	1	2	3
Electricity – lighting	All lights off when shutdown except for emergency and pathway lighting				
Electricity – lighting	Changed lighting system to environmental friendly bulbs			X	
Electricity – lighting	High efficiency lighting in sheeter, finishing, receiving and warehouse	complete			
Electricity – lighting	High efficiency light project	in process			X

CURRENT PRACTICES			Plant		
Area	**Sustainability Practice**	**Comments**	**1**	**2**	**3**
Electricity – lighting	Installed automatic lights in cafeteria and upstairs bathroom				
HVAC	Turn off MFG Equipment when not in use (knives, joggers, shrink wrap machines)				
Non-process	Programmable thermostats in many areas	more area to go			
Recycling – customer	Using recycled paper cups & paper towels				
VOCs	Light bulbs, batteries, ballasts recycled				
Waste – other	Maintain RTO system efficiently			X	
Waste – process	Recycle lightbulbs				X
Waste – process	Print on backside of sheet during make ready				
Waste – process	Recycling Corrugate		X		
Waste – process	Reduced roll size on the seamer from 930 meters to 905 meters to reduce seamer waste	Investment in new seamer		X	
Waste – process	Reuse cores from press and back again				X
Waste – process	Reuse corner boards from incoming synthetic paper to reuse for shipping finished products				
Water – non-process	Automatic flush valves on washroom toilets and urinals		X		
Water – process	Recycle water from platemaking				
PROPOSED PRACTICES			Plant		
Area	**Sustainability Practice**	**Comments**	**1**	**2**	**3**
Electricity – non-process	Remove bulbs and ballasts where not needed; use LED or CFL or equivalent lighting and occupancy sensors	Low cost energy reduction suggestions			
Electricity – non-process	Turn off PCs at end of day				
Electricity – process	Additional Light switches in areas that are not in use				
Electricity/ Natural Gas	Increase thermostat temp settings in summer and decrease in winters; and reduce hot water heater temp to 115 F	Low cost energy reduction suggestions			

PROPOSED PRACTICES			Plant		
Area	Sustainability Practice	Comments	1	2	3
Electricity/ Natural Gas	Keeping doors closed to prevent loss of cooled air from the AC; decrease utility cost				
Electricity/ Natural Gas	See about insulating above the drop ceiling for more energy savings				
Electricity/ Natural Gas	Use programmable thermostats to adjust temps in office areas during afterhours/ weekends		X		
Other Miscellaneous					
Non-process waste	Non-process waste recycling	Need Definition of non-process waste			

As a result of monitoring the use of electricity, water, gas, paper and chemicals, the company I worked for was able to make several improvements by putting together a cross-functional team on sustainability and sustainable business practices. The goal of the team was to reduce, reuse, and recycle hazardous and non-hazardous process waste as best as possible.

For example, one of the plants had a much higher rate of gas usage one month, even though there was no significant change in operation time or weather data. After a detailed evaluation of the plant, a gas leak was found in the supply pipe outside the plant. What if the gas leak had been inside the plant? Not only was the gas leak fixed, but the plant's overall safety was also improved by installing a gas detection system throughout the plant.

Similarly, when I was reviewing the VOC emitted from two similar plants, I noticed that one plant had reported a significantly higher amount of VOC than the other to the EPA. The level of VOC emissions permissible at plant one, according to the EPA, was higher than the level permissible at plant two because plant one was located in a rural area while plant two was located in an urban area. So plant one was using chemicals that emitted higher VOC levels because it was monitoring the total amount of VOC and keeping it under the permissible limits allowed by the EPA. We decided to change the chemicals being used here, even though plant one was legally allowed to emit higher levels of VOC because it was the right thing to do. We reduced the VOC emitted to the atmosphere using low-VOC chemicals. That is what world-class safety companies do.

Another issue came up with water usage. One of the plants was using a significant amount of water to dilute the hazardous waste generated in the manufacturing process and then dumping it into the sewer system. Since the PH level of the diluted water was low, it was technically permissible for a Publicly Owned Treatment Works (POTW). In contrast, another facility treated the waste so that it could be recycled and reused. However, it required special equipment to accomplish this. When ROI calculations were completed around how different facilities dealt with waste, it became clear that the first plant should switch to the second plant's process and they made the necessary investment. This change was initially difficult for some people because they had been doing it this way for years. But they couldn't argue with the fact that a lot of money had been wasted over the years and, at the same time, their dilution method was reducing the treatment capacity of the POTW.

The bottom line is that world-class safety companies go above and beyond just protecting their employees and facilities. They make every effort to protect the environment and the community in which they operate.

At one specific printing company, waste reduction and recycling became the most important driver for improving their environmental footprint. By minimizing waste in its manufacturing plants, the company was able to reduce its use of electricity, natural gas, and water and VOC emissions. Waste minimization was targeted through effective engineering and focused efforts at preparing each job efficiently. The company was able to achieve its goals by creating a comprehensive policy statement similar to the one outlined below.

Sustainability Position Statement

Our company is committed to the support of sustainable business practices that recognize our social responsibility, minimize our environmental footprint, and offer customers innovative product solutions to help them achieve their goals. Continuous improvement of our efforts will provide enhanced competitive advantage, lower operating costs, and continued customer value.

Our focus will be on areas that have the most impact and are supportive of our core strategies of customer intimacy, financial discipline, and a performance driven culture. To measure the success of our programs and communicate our progress, we have developed goals and key metrics for driving results. These metrics will also be used to provide meaningful and relevant data for analyzing trends and for identifying additional opportunities for improvement.

Sustainability goals include:

- *Waste management:* Minimize internal waste produced while increasing the recycling and/or reuse rate for any waste generated
- *Water management:* Effectively manage the consumption and disposal of water used in our manufacturing processes
- *Energy efficiency:* Conserve resources and energy usage in our manufacturing processes and buildings (including natural gas, electricity, and heating oil) through effective design and evaluation criteria for new and existing assets
- *Emissions:* Meet all regulatory and governmental air quality regulations while striving to exceed required standards
- *Transportation:* Leverage our geographic footprint to utilize the most effective and efficient methods for servicing our customers; utilize locally sourced raw materials and supplies whenever practical
- *Sustainable product design:* Product design development and sourcing based upon established sustainable product design criteria

Product design criteria include:

- To the extent practical, our suppliers should provide raw materials that are beneficial, safe, and healthy for individuals and communities throughout their life cycles
- Meets criteria for performance and cost
- To the extent possible, we will optimize use of alternative and renewable energy in sourcing, manufacturing, transportation, and recycling of labels
- Maximizes the use of renewable or recycled source materials
- Use clean production technologies and best environmental practices

Our sustainability goals can be achieved by monitoring the results of water, electricity, and gas usage and developing various sustainability initiatives that include water consumption, recycling and waste reduction. It always helps to discuss success stories, and monthly conference calls with sustainability champions can help in monitoring the success of the program. A plan to monitor and reduce water usage requires:

1. Completing a process flow diagram showing how the water is routed and used through the plant, including where it came from (well, city main, etc.) and where it is disposed (such as sewer, recycled, reused, etc.)
2. Describing the process where water was used in the plant such as boiler, pre-press, press room fountain solution, office, etc.
3. Estimating the amount of water used in each of the processes identified above in (b).
4. How to reduce the water usage by 10 percent?

Samples of a water-usage report and process-flow diagram are included here as well.

Water and Sewer Usage

Plant Number	Bath Rooms	Production	Cooling System	Landscape	Total Monthly Estimated Gallons	Recommendation to Reduce, Reuse, Recycle
1	44,000	32,000	14,000		90,000	Water leak in the Litho Press Department accounted for 65% of total waste. This was discovered and rectified with a total savings of $4,500. 2. Savings could be obtained from replacing with updated lower volume toilets
3	60,000				60,000	Shut off corporate side heating and cooling to save water and other utility. 2. Replace existing valves on toilets with water saving units
4	11,250	11,500	4,250		27,000	10% savings in water usage from better wash-out system installed recently. Generally low demand water system is in place
5	10,000				10,000	Water used for sanitary purpose and press fountain solution
6		15,000				Turn water off in pre-press when not needed. 2. Will install fountain solution filter system to reduce water waste

So that corporate can successfully manage the hazardous waste generated at each plant, it is important that the environmental safety leader and other managers take proactive measures to reduce hazardous waste and monitor disposal of any hazardous waste generated. It is important to know the site specific requirements to properly dispose of the waste. This program is an attempt to set the tone and provide guidance to comply with the Resource Conservation and Recovery Act (RCRA) and local standards.

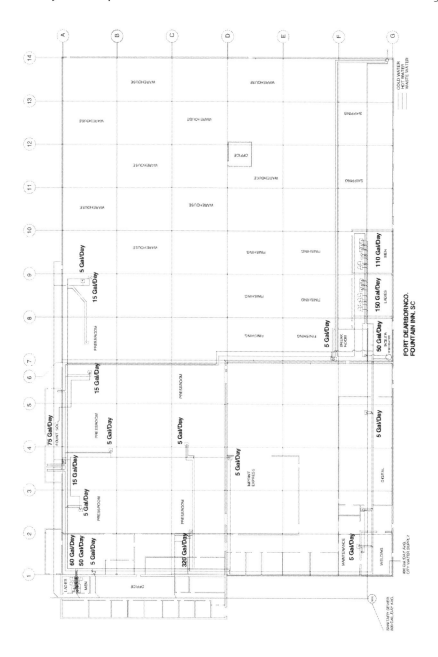

On the next pages I have listed a method that can be used to monitor hazardous and non-hazardous waste and that will ensure it is properly disposed of. At the same time, I also provide insights into how waste can be recycled or minimized.

Hazardous and Non Hazardous Waste Hauling and Disposal

Identity Profile	Type of Waste	Quantity	Disposal Method	Manifest No.	Date Transported	Disposer	Hazard
12-024-03	UV Ink	4105 lb	Thermal Destruction	3233232	1/11/08	Emco Waste Services	Non-Hazardous
	Lamps	729 each	Recycle	BOL 3710	1/23/08	Mercury Waste Services	Universal Waste
12-024-03	UV Ink	3706 lb	Thermal Destruction	3233359	1/25/08	Emco Waste Services	Non-Hazardous
12-024-03	UV Ink	5245 lb	Thermal Destruction	3233495	2/22/08	Emco Waste Services	Non-Hazardous
12-024-03	UV Ink	3765	Thermal Destruction	3233409	2/8/08	Emco Waste Services	Non-Hazardous
12-024-03	UV Ink	3809 lb	Thermal Destruction	3233357	3/10/08	Emco Waste Services	Non-Hazardous
03-021-08	Bio Diesel	315 lb	Thermal Destruction	3835613	3/20/08	Emco Waste Services	Non-Hazardous
12-024-03	UV Ink	3665 lb	Thermal Destruction	3835613	3/20/08	Emco Waste Services	Non-Hazardous
12-024-03	UV Ink	1630 lb	Thermal Destruction	3835647	3/28/08	Emco Waste Services	Non-Hazardous
12-024-03	UV Ink	4200 lb	Thermal Destruction	3835712	4/11/08	Emco Waste Services	Non-Hazardous
12-024-03	UV Ink	5283 lb	Thermal Destruction	3835781	4/25/08	Emco Waste Services	Non-Hazardous
12-024-03	UV Ink	4110 lb	Thermal Destruction	3835812	5/9/08	Emco Waste Services	Non-Hazardous

12-024-03	UV Ink	2180 lb	Thermal Destruction	3835946	5/30/08	Emco Waste Services	Non-Hazardous
12-024-03	UV Ink	2170 lb	Thermal Destruction	3835975	6/6/08	Emco Waste Services	Non-Hazardous
12-024-03	UV Ink	4315 lb	Thermal Destruction	3835914	5/23/08	Emco Waste Services	Non-Hazardous
12-024-03	UV Ink	4210 lb	Thermal Destruction	3836519	6/20/08	Emco Waste Services	Non-Hazardous
12-024-03	UV Ink	4080 lb	Thermal Destruction	3836627	7/11/08	Emco Waste Services	Non-Hazardous
12-024-03	UV Ink	3100 lb	Thermal Destruction	3836580	6/27/08	Emco Waste Services	Non-Hazardous
12-024-03	UV Ink	3050 lb	Thermal Destruction	3836654	7/25/08	Emco Waste Services	Non-Hazardous
12-024-03	UV Ink	3250 lb	Thermal Destruction	3836736	8/8/08	Emco Waste Services	Non-Hazardous
12-024-03	UV Ink	4190 lb	Thermal Destruction	3836800	8/25/08	Emco Waste Services	Non-Hazardous
12-024-03	UV Ink	3245 lb	Thermal Destruction	3836871	9/5/08	Emco Waste Services	Non-Hazardous
	Normal Propyl Alcohol and Propyl Acetate	1 drum	Fuel Blending	3836802	8/25/08	Emco Waste Services/ Badger Disposal	Hazardous
12-024-03	UV Ink	3075 lb	Thermal Destruction	3836923	9/19/08	Emco Waste Services	Non-Hazardous
12-024-03	UV Ink	2150 lb	Thermal Destruction	3836956	9/29/08	Emco Waste Services	Non-Hazardous

Like I said earlier, companies with world-class safety culture go beyond just protecting employees from work related hazards. They make their best efforts at protecting the environment as well.

Business Continuity Plan

FDC has developed a Business Continuity Plan (BCP) for effectively responding to various disasters that could significantly disrupt their business and the businesses of their customers. The objective of this plan is to minimize, if not eliminate, interruptions to critical business functions as a result of catastrophic, unforeseen events. The BCP plan includes:

- Corporate and Plant Structures & Responsibilities
- Risk Assessment and Business Impact Analysis
- Business Continuity Plan Activation Procedures
- Notification and Communication Procedures
- IT Disaster Recovery Plan
- Plant Operations Continuity Plan
- System Data and File Integrity Plan
- Finance and Human Resources Plan
- Supplier Contingency Plan

Our objective in developing the Business Continuity Plan is to implement appropriate procedures so as to ensure associates' health and safety, protect customer interests, maintain the ability to conduct critical business functions, ensure appropriate communications to associates and business partners, minimize damage to company assets and infrastructure, and implement a disaster-recovery plan.

In order to accomplish these goals, the plan requires the involvement and engagement of various associates who are authorized and have the skills to deal with specific emergencies. The executive management team should provide necessary authorizations to other teams, declare disaster (if called for), and lead important communications. The BCP core team, which includes senior level directors, provides vital support and coordination functions to the plants. The plant emergency response team provides the first response in emergency situations and is responsible for assessing the extent of the damage, preventing further damage, and completing restoration activities.

We understand that communication is critical when a disaster strikes any of our facilities. Currently, we communicate with our associates, customers, and vendors via a variety of communication methods. In the event of a business disruption, we assess which means of communication are still available and utilize proper methods to quickly respond to the emergency. One or all of the following methods may be utilized for communicating with associates and customers during and after a business disruption situation:

- BCP Intranet site
- Face-to-face meetings
- Voicemail or email
- Toll-free hotline
- Call tree

Our disaster communication procedures include associate notification, customer notification, supplier notification, and media and public notification procedures.

Technology Disaster Recovery Plan

We have developed a comprehensive Technology Disaster Recovery Plan that includes:

- Critical IT Services & Supporting Systems
- Risk Mitigation & Disaster Prevention
- Tape Backup & Data Mirroring
- Operating & Contingency Procedures
- System Administration Practices
- System Restoration Procedures
- Use of alternate disaster recovery site
- Communication Plan
- Plan Testing

Plant-specific Disaster Recovery Plan

Each of our facilities has created its own Plant-specific Emergency Action Plan which is maintained under the same guidelines as the Corporate Master plan. This detailed action plan is used as a training and reference guide for the facility associates. The plan includes such things as how to prepare for an emergency. It defines roles and responsibilities for events such as fire, chemical spills, bomb threats, hostile intruders, and other emergencies. The plan includes plant-specific emergency contacts and instructions on how to manage the emergency. Site-specific maps that include such items as emergency exits, fire extinguisher locations, fire sprinkler zone valve locations, and all utility shut-off locations are an integral part of the plant's BCP. We also have in place a Business Transfer Plan to transfer business from an affected facility to one of our other facilities.

Human Resources Continuity Plan

The Human Resources Continuity Plan covers:

- Pandemic Outbreak
- Hostage Situation

- Payroll Processing
- Workers Compensation
- Benefit Administration
- Personnel File management

Supplier Continuity Plan

The corporate director of purchasing performs a risk assessment on corporate suppliers. All suppliers of critical materials are asked to provide a copy of their Business Continuity Plan. These plans are used to evaluate supplier controls in case of an emergency and to decide whether to seek alternate suppliers for the materials to meet our customer needs.

FDC utilizes several different sources for their raw materials. Each of our corporate suppliers has multiple manufacturing plants for the products we purchase. We have established back up manufacturing plants that can be used in case of a disaster at the primary facility

To ensure that the Business Continuity Plan can be successfully implemented when needed, associates, including management, should be regularly trained in emergency and business continuity procedures. A full comprehensive test of the entire plan is not feasible or practical; however, individual portions of the plan and procedures should be tested on a regular basis to meet company needs.

Training

To ensure that the Business Continuity Plan can be successfully implemented, the company's corporate and plant management and other key associates should be trained in emergency and continuity procedures. These training programs include any or all of the following:

a. A PowerPoint training presentation covering the company's Business Continuity Plan and its goals, objectives, and procedures to assist in the management of any disaster.

b. Training programs from Coastal, HI 5 Productions, Summit, and other training media to increase awareness among the management staff and assist in the prompt and correct response to a disaster. For this purpose, we use the training program titled, "High Impact Emergency Action Plans" by HI 5 Productions

c. Training in how to respond to various potential scenarios such as a power outage, flood, tornados, fire, etc.

It is the responsibility of the BCP core team and plant general managers to develop and conduct appropriate training on a regular basis.

Plan Testing

The BCP plan should be tested by using internal and external resources. A full comprehensive test of the entire plan is not feasible. Hence, individual portions of the plan and procedures should be tested and verified on a regular basis. Every test should be documented and have its results analyzed. The goal of the tests is to find gaps so as to improve the plan.

The internal test is a review of the response to various scenarios presented to the management group being tested. The external test may be completed by the Building Operation Management Association (BOMA) or any other similar organization. The BCP core team is responsible for defining, scheduling, performing, and reviewing test results.

As stated earlier, it is important to test the BCP plan. We used scenario testing to test various possible scenarios. One such scenario is as follows:

At about 10 p.m. on Saturday, one crew had completed a job and had just set the press up for water-based coating. This setup required them to increase the temperature of the air knives to about 180 degrees. The crew had just started to run a job when the fire started. One of the operators noticed smoke and fire coming from the hoses leading to the knives. The smoke and flames spread between the two cylinders.

What do you do next?

<u>Record Your Actions Here</u>

It is important to keep minutes of all BCP-related planning meetings so that progress can be reviewed and action can be taken in developing an excellent plan. Below is a sample copy of the minutes from one of our BCP core team meetings.

Meeting Date: 6/10
Meeting Time: 2 p.m. CST
Attendees:

Agenda

Item	Responsible	Time
Review progress made	Bill	
Review master plan outline	Mike	
Review identified gaps in the plan	All Team Members	
Formulate actions and objectives	All Team Members	

Bill was disappointed with the progress we have made so far particularly since we plan to complete BCP document by the end of the year. He felt we all need to focus and step up our efforts a notch.

This is important particularly since some of our large customers expect us to have a viable Business Continuity Plan in place.

Bill would like us to have two documents ready by the end of the year: External BCP Document and an Internal BCP Document. The external document would be an overview of our plan and accessible to all of our customers. The internal document would show a Road Map for our success in managing our business in case of a disaster. This document could be an electronic BCP folder.

Mike then reviewed what we have been able to accomplish so far this year:

a. We have developed the BCP core team to assist management through a disaster.

b. We have completed BCP workshops to make senior managers aware of the need to be prepared for a disaster.

c. We have developed the Event Classification Matrix to identify the responses from each of the three categories of events.

d. We have developed the FDC Capability Matrix, a matrix that makes it easy to leverage our different plants in case of a disaster.

e. We have developed the Business Continuity Phases to assist managers in taking appropriate actions at different phases of a disaster.

f. We have developed an electronic file folder where all team members can review the program.

g. We have developed an outline and framework for organizing the BCP as reflected under the table of contents

h. We have developed the Insurance Information to report a disaster and ask for insurance company assistance.

i. We have developed Emergency Action Plans for each of our facilities.

j. We have developed the Planning Process for Transfer of Work Between Plants.

k. We have developed the Supplier Continuity Plan to make sure the plants receive the raw materials they need to produce the labels. Obviously this is an ongoing process.

l.   We have completed a Technology Disaster Recovery Plan.

As you all can see, we have accomplished a lot within a short six-month period. What we need to do in the future is to finish the assignment each one of us has undertaken. In particular, we need to get the following accomplished:

1.   Finance Continuity Plan
2.   Human Resource Continuity Plan
3.   Vital Records Protection Plan
4.   Facility and Asset Restoration Plan

We agreed to meet monthly to review the progress made towards accomplishing our goal of completing the plan by the end of the year so we can test the plan early next year.

Action Item	Responsible	Due Date
Finance Continuity Plan	F	
Human Resource Continuity Plan	H	
Vital Records Protection Plan	V	
Facility and Asset Restoration Plan	D	

Completed Action Items

Action Item	Responsible	Date Completed
As documented above		

The Business Continuity Plan and several other similar programs were not required by OSHA or other government standards. However, FDC, being a world-class company, made sure its management team spent the time and resources necessary to put in place these programs to protect its employees, customers, and operations because it was the right thing to do.

Defensive Driving

World-class safety companies promote safe driving practices for their employees, which benefits the company and employee families. Some of the guidelines developed include:

Vehicle Usage Policy:

The purpose of this policy is to ensure the safety of those individuals who drive on company business using their personal vehicles, rented vehicles or company owned vehicles. The goal is to offer guidance on the proper use of vehicles since vehicle accidents are costly; but more importantly, they may result in injury to you or others. It is your responsibility to operate the vehicle in a safe manner and to drive defensively to prevent injuries and property damage. The attitude you take when behind the wheel is the single most important factor in driving safely. Each associate is expected to drive in a safe and courteous manner pursuant to the state and local laws and safety rules.

The Fleet Accident Review Committee is comprised of:

The corporate safety director and the human resources director and the direct report for the associate involved in an automobile accident. The committee will be responsible for:

1. Reviewing accidents and overall driver safety record for associates assigned to it to determine if there should be changes in policy or procedure; or if other corrective action (such as training, equipment changes, etc.) should be implemented to enhance the safe operation of vehicles and employee safety on

company business and personal business on company owned vehicles.

2. With guidance from the human resources director reviewing driving records of individual associates reporting to them and making recommendations if and when driving privileges should be suspended or revoked.

3. Reviewing all other issues that arise with respect to compliance with this policy.

Driver Guidelines and Reporting Requirements:

1. Company owned or leased (includes rental) vehicles are to be driven by authorized employees only.

2. Any employee who has a driver's license revoked or suspended shall immediately notify his manager, and **immediately discontinue operation of that vehicle on company business**. Failure to do so may result in disciplinary action, including termination of employment.

3. All accidents while on company business, regardless of severity, **must** be reported to the police and to your manager. Accidents are to be reported immediately as practicable. Failing to stop after an accident and/or failure to report an accident may result in disciplinary action, up to and including termination.

4. Drivers **must** report to their manager all traffic ticket violations received while on company business driving personal vehicles, rented vehicles or company-owned vehicles.

Driver Criteria, Administration, and Verification:

Associates who will be using their personal vehicle while conducting company business or using company owned or leased vehicles **must** provide a valid and current driver's license and insurance card to human resources at the time of hire. Additionally, Motor Vehicle Records (MVR) will be run by the human resources department with the assistance from the insurance agency at the time of hire, and preferably annually each year thereafter.

Associates are expected to drive in a safe and responsible manner and to maintain a good driving record. The Fleet Safety Committee is responsible for reviewing records, including accidents, moving violations, etc., to determine if an employee's driving record indicates a pattern of unsafe or irresponsible driving, and to make a recommendation for changing this unsafe driving or suspension/revocation of driving privileges.

Criteria that may indicate an unacceptable record includes, but is not limited to:

1. Three or more moving violations in a year while on company business.
2. Three or more chargeable accidents within a year while on company business. Chargeable means that the driver is determined to be the primary cause of the accident through speeding, inattention, etc.
3. Any combination of three accidents and/or moving violations.

Driver Safety Rules:

1. The use of a vehicle on company business while under the influence of intoxicants and other drugs over the state permissible limit which could impair driving ability is forbidden and is sufficient cause for discipline, up to and including termination.
2. For your safety, cell phone use while driving is not permitted. While driving, attention to the road and safety should always take precedence over conducting business over the phone.
3. No driver shall operate a vehicle on company business when his/her ability to do so safely has been impaired.
4. All drivers and passengers operating or riding in a vehicle on company business **must** wear seat belts as is required by the state laws even if air bags are available.
5. All state and local laws must be obeyed.

Accident Procedures:

1. In an attempt to minimize the results of an accident, the driver must prevent further damages or injuries and obtain all pertinent information and report it accurately.
    - Call for medical aid if necessary.
    - Call the police. All accidents, regardless of severity, must be reported to the police.
    - Record names and addresses of drivers, witnesses, and occupants of the other vehicles and any medical personnel who may arrive at the scene.
    - Complete the form located in the vehicle accident packet. Pertinent information to obtain includes: license number of other drivers; insurance company names and policy numbers of other vehicles; make, model, and year of other vehicles; date and time of accident; and overall road and weather conditions. If possible, take pictures of the scene of the accident.
2. Do not discuss the accident with anyone at the scene except the police. Do not accept any responsibility for the accident. Don't argue with anyone.
3. Provide the other party with your name, address, driver's license number, and insurance information.
4. Immediately report the accident to your manager. Provide a copy of the accident report and/or your written description of the accident to the manager and the Fleet Safety Committee.
5. There will be an accident review conducted by the Fleet Accident Review Committee on each accident to determine cause and how it could be prevented in future. If the accident is found to be chargeable, at fault, then the driver will be responsible for half of the deductible but no more than $250.

The goal of the safe driving policy and program is to make sure employees drive personal and company vehicles safely for the sake of the company and their families.

Safety at Home

World-class safety companies emphasize the importance of safety at home. Whether you have an accident at home or at work, you will not be able to perform your work responsibilities at full capacity if you are injured. Companies understand that. And they work on making sure the employees understand that as well. When I was first employed at American Mutual Insurance Company (many of you may not remember that name), my wife was pregnant with our first child. I could not focus my mind on work while I was worrying about how she would get to the hospital when the time came to deliver the baby. My boss, a burly old man originally from South Carolina, told me to go home and take care of my family. If your child gets hurt at home because someone did not follow safety precautions—such as leaving toys on the stairs that could cause the little one to come tumbling down—you will be worrying at work, thinking about how he is recovering. That distraction can easily cause an injury at work to you or your fellow workers. World-class safety companies understand that, which is why they teach people to be safe at home and train their family members about potential hazards.

One of our managers hurt himself using a chainsaw to cut wood. At our safety meetings we discussed various safety precautions one should take while operating power tools such as chainsaws, table saws, miter saws, and other common household power tools. The precautions we discussed include:

- Read and familiarize yourself with the manufacturer's instructions, including precautions and how to respond to an emergency.
- Use recommended protective equipment, such as safety goggles, clothing, earplugs, dust masks, and gloves.
- Make sure that the tool is in good working order, including the cord, switch, and any protective guards.

- Never modify a tool for a job it's not intended to do.
- Keep safety switches in working order; do not bypass or replace them with standard switches.
- Avoid accidental starts by keeping hands away from switches while carrying plugged-in or battery-powered tools.
- Avoid using tools if you are taking strong medications or consuming alcohol.
- Keep work areas well lit and free of clutter. Never use power tools in damp or wet locations.
- Have observers remain a safe distance away from the work area.
- Keep good footing and maintain good balance.
- Avoid loose clothes, ties or jewelry.
- Store tools in a dry place that is not above your head, out of the reach of children, and disconnect the power supply.

One of our employees brought in a news clipping showing how a someone cutting small bushes around the house got hurt by using the lawnmower improperly to trim the bush. He ended up cutting the end of his toes instead of the bush! We discussed lawnmower safety at several of our safety meetings after that. Some of the safety precautions we discussed as provided by the manufacturer of lawn mowers, Cub Cadet include:

Preparing to use a push lawn mower:

- When using a push lawn mower, always wear safety glasses or safety goggles. This lawn mower safety tip should be followed even while performing an adjustment or repair. Objects thrown from a push lawn mower can ricochet and cause serious injury to the eyes.
- When using a push lawn mower, wear sturdy, rough-soled work shoes (preferably steel-toed) and close-fitting slacks and shirts that cover the arms and legs. Never operate a push lawn mower with bare feet, sandals, slippery or light-weight shoes.

- Before you power up your push lawn mower:
- Take a short walk around your lawn and thoroughly inspect the area where the push lawn mower will be used. Keeping lawn mower safety in mind, remove all stones, sticks, wire, bones, toys, and other debris from the lawn that could be tripped over or picked up and thrown by the blade.
- Plan your mowing pattern. Be aware that your push lawn mower can discard materials as you move. For lawn mower safety, avoid roads, sidewalks and bystanders, as well as walls or obstructions, which may cause discharged material to ricochet back toward you.
- Be sure to inspect the push lawn mower that you plan to use. Keep an eye out for damaged or missing parts and make any adjustments or repairs before you begin. Missing or damaged parts can cause blade contact or thrown object injuries.
- When your push lawn mower is in motion, use these safety tips:
- If lawn mower safety is your priority, remember that many injuries occur as a result of the push lawn mower being pulled over the foot during a fall. Do not hold on to the push lawn mower if you are falling; release the handle immediately.

When walking with a push lawn mower, never pull the mower back toward you while moving. If you must back the push lawn mower away, remember these lawn mower safety tips:

- First, look down and behind you to avoid tripping.
- Step back from the push lawn mower to fully extend your arms.
- Be sure you are well balanced with sure footing.
- Pull the lawn mower back slowly, no more than half way toward you.

We discussed the importance of having a fire evacuation plan for employee residences and arranging regular drills where parents can take the lead in preparing their children. We discussed the typical fire evacuation plan supplied by American Red Cross and NFPA, which includes the following advice:

- Install a smoke alarm on every level of your home and outside of sleeping areas.
- Test smoke alarm batteries every month and chang them at least once a year.
- Make sure everyone knows at least two ways to escape from every room of the home.
- Practice your fire escape plan at least twice a year. Designate a meeting spot outside and a safe distance from home. Make sure all family members know the meeting spot.
- Have family practice escaping from home and low crawling at different times of the day. Make sure everyone knows how to call 9-1-1.
- Consider escape ladders for sleeping areas on the second or third floors. Make sure everyone in your home learns how to use them ahead of time by reading the manufacturer's instructions and understanding the steps to use them. Store them near the window where they will be used.
- Teach family to stop, drop to the ground, and roll if their clothes catch on fire. Practice this with your children.
- Once you get out of your home, stay out under all circumstances until a fire official gives you permission to go back inside.
- Never open doors that are warm to the touch.
- If smoke, heat, or flames block your exit routes, stay in the room with the door closed. If possible, place a towel under the door and call the fire department to alert them to your location in the home. Go to the window and signal for help waving a bright-colored cloth or a flashlight. Do not break the window, but open it from the top and bottom.
- Visit www.redcross.org/homefires for more information on creating home fire escape plans.

We also discussed what the employees can do to prevent falls at home. According to a recent survey by the Home Safety Council, nearly 5.1 million Americans are injured each year from falls in and around the home. Children under five and adults over sixty-five are at greatest risk of fall-related injuries at home, but according to the Consumer Product Safety Commission, falls are the leading cause of nonfatal injury for all age groups except those fifteen to twenty-four years of age. "Our research confirms that most Americans do not realize that falls are by far the most common cause of unintentional injury within the home," said Home Safety Council President Meri-K Appy. To keep your employees safe from the risk of a fall at home, the council offers these tips:

- All stairs and steps should be protected with a secure banister or handrail on each side that extends the full length of the stairs. Make sure stairwells have a bright light at the top and bottom of the stairs.
- Make sure all porches, hallways, and stairwells are well lit. Use the maximum safe wattage in light fixtures. Maximum wattage is typically posted inside light fixtures.
- Use nightlights to help light hallways, stairwells, and bathrooms during night-time hours.
- Keep stairs, steps, landings, and all floors clear. Reduce clutter and safely tuck away telephone and electrical cords out of walkways.
- In homes with children, make sure toys and games are not left on steps or landings.
- When very young children are present, use safety gates at the tops and bottoms of stairs.
- Use a non-slip mat or install adhesive safety strips or decals in bathtubs and showers. If you use a bath mat on the floor, choose one that has a non-skid bottom.
- Install grab bars in bath and shower stalls. Do not use towel racks or wall-mounted soap dishes as grab bars—they can easily come loose, causing a fall.

- Keep the floor clean. Promptly clean up grease, water, and other spills.
- If you use throw rugs in your home, place them over a rug-liner or choose rugs with non-skid backs to reduce your chance of slipping.
- Know that window screens are not strong enough to protect a child from falling out. Install window guards on upper floors, making sure they're designed to open quickly from the inside in case of fire.
- Always practice constant supervision if children are near an open window, and keep cribs and furniture away from windows.
- Follow medication dosages closely. Using multiple medications and/or using medications incorrectly may cause dizziness, weakness, and other side effects which can lead to a dangerous fall.
- On a playground, cover areas under and around play equipment with soft materials such as hardwood chips, mulch, shredded rubber, pea gravel, and sand. Materials should be 9 to 12 inches deep and extend 6 feet from all sides of play equipment.
- When climbing on a ladder is necessary, always stand at or below the highest safe standing level. For a stepladder, the safe standing level is the second rung from the top. For an extension ladder, it's the fourth rung from the top.

An aging workforce reinforces the need for off-the-job safety as employee injuries prove hazardous to employers' bottom lines.

In today's workforce, employers should be happy to know that they can continue to depend on a mature and experienced pool of employees. According to the U.S. Bureau of Labor Statistics, workers fifty-five and older represent the fastest growing segment of the workforce, approximately 22.7 million.

According to a recent study conducted by the Home Safety Council, unintentional home injuries cause nearly 20,000 deaths and 21 million medical visits on average each year, with the highest rate of unintentional home injuries and injury-related death coming from the older adult population. It's no secret to world-class employers that a healthy employee equals a healthy bottom line, but employers may not realize what home injuries can do to the bottom line. The answer is about $38 billion a year.

Employers are spending an average of $280 per employee equating to $38 billion annually due to home hazards including falls, fires and burns, poisonings, suffocation and drowning. Health insurance, life insurance, sick leave and disability, hiring and training new employees, not to mention the time-off employees may need to care for a loved one who has sustained a debilitating home injury, all hit employers in the pocket book.

By understanding that one-third of all home injury-related deaths and nearly 2.3 million unintentional home injuries occur among the older adult population, and knowing that this age group makes up nearly half of the U.S. workforce, more employers should consider these figures a call to action.

Employer Action

While most businesses already invest in the safety and well-being of their employees at work, there are simple steps that can extend these benefits back to the employee's homes and families. Steps employers can take to promote home safety practices among their employees, young and old, can complement workplace safety programs already in place, such as:

- Include a home safety section in company newsletters
- Host quarterly safety fairs
- Insert a home safety agenda item during employee staff meetings
- Organize weekly home safety huddles

These and many more creative and effective home safety education efforts will help employers not only increase the safety and wellness of all their workers, but also reduce the financial toll unintentional home injuries can take on the business.

One home safety resource available for employers is the Home Safety Council Web site at www.homesafetycouncil.org.

Indoor Air Quality

World-class safety companies provide good indoor air quality to their employees and complete IH audits to make sure the air quality is the best.

Some of the indoor air quality issues that world-class safety companies should be evaluating and managing include:

- Radon
- Secondhand smoke
- Mold and other allergens
- Carbon monoxide
- Various types of bacteria such as legionella
- Asbestos
- Carbon dioxide
- Ozone pollution
- Volatile organic compounds

Discussion of each of these indoor air quality issue is outside the scope of this book.

World-class safety companies think outside the box when it comes to ensuring the safety of their employees because healthy and happy employees benefit the company and society.

Acknowledgments

First of all, I would like to thank my daughter, Reshma Saujani, who has been an inspiration, particularly with her book *Women Who Don't Wait in Line*. Her book inspired me to get up every day, forget the past, live in the present, and do something good for the company employees. I would like to thank her for finding Katie Salisbury, the editor who worked with me on this book. She has been just outstanding. Her careful comments and criticism—*maybe you want to explore a little bit more here* or a comment like *you may want to be a little more clear here*—helped me make the contents of the book clear for all readers.

I would like to thank my colleagues and employees, particularly the safety and HR team leaders at Fort Dearborn Company for applying these safety principles, sometimes with open arms and sometimes begrudgingly, with outstanding results for themselves and the company. In particular, I would like to thank Bill Johnstone, COO at FDC, whom I reported for his guidance, support, and insight. I would like to thank Mike Anderson, CEO at FDC, for his generous bonus and for recognizing safety as a critical business function.

I would like to thank the wonderful folks Brett Carroll, loss-control account manager; Rex DePeel, executive consultant at The Hartford; Kenneth White, risk control consulting director, and Lawrence Skalnik, equipment breakdown consulting director at CNA Insurance Companies for their assistance and advice in making safety and loss prevention ideas work at Fort Dearborn Company. I would like to thank Anthony Crissie of Crissie Insurance Group for his insurance and risk management insights.

This book project became to life after the publication of two of my articles—one in *The Insight*, a publication of the Charter Property and Casualty Underwriters Society (CPCU), and the other in ASSE's journal *Professional Safety*. I would like to thank John Kelly, director of content acquisition at CPCU, and Sue Trebswether, editor at ASSE, for their assistance and inspiration in getting those articles published.

And, finally, I would like to thank my wife, Meena Saujani, for bearing with me sitting in front of my computer, plugging away at page after page instead of helping her with normal household chores.

I hope the book inspires you to do something innovative to promote safety and the well-being of the employees you are responsible for.

Bibliography

"ABCs of Experience Rating." National Council on Compensation Insurance (NCCI). Accessed April 30, 2015. https://www.ncci.com/Articles/Documents/UW_ABC_Exp_Rating.pdf

Campbell Institute. "Defining World-class EHS: An Analysis of Leading EHS Management System Practices of Robert W. Campbell Award Winners." National Safety Council. Chicago, Illinois, 2012.

Eckenfelder, Donald J. "Getting the Safety Culture Right." *EHS Today*, October 15, 2003. http://www.ehstoday.com/mag/ehs_imp_36651

Flin, R. and S. Yule. "Leadership for Safety: Industrial Experience." *Quality Saf Health Care,* 2004. http://qualitysafety.bmj.com/content/13/suppl_2/ii45.full

Hansell, Cathy A. "Become a Transformational Safety Leader." Presentation at America's Safety Conference, Chicago, Illinois, September 2012.

Hansell, Cathy A. "Successfully Aligning and Integrating Safety (SH&E) Within the Business." Breakthrough Results, LLC, North Venice, Florida. http://www.asse.org/assets/1/7/601.pdf

Heitzberg, J. "Process Control Chart Tool Kit." *IBM Reference Manual for Windows*, version 6.0.2., 1985-1997.

Jones, David A. "Mapping Support for an EHS Management System." *Occupational Hazards,* June 29, 2006.

Middlesworth, Mark. "A Short Guide to Leading and Lagging Indicators of Safety Performance." *Ergonomics Plus*, April 24, 2013. http://ergo-plus.com/leading-lagging-indicators-safety-preformance/

Murali, J. "Roadmap to World-class Safety for L&T." Presentation at HCP, NRO – New Delhi conference, New Delhi, India, January 31, 2012.

Pais, David S. "Showing EHS Value Through Return on Investment (ROI)." Presentation at the ASSE Professional Development Conference and Exposition Safety, Chicago, Illinois, June 15, 2011.

Saujani, Michael and Nick Adler. "Safety at Fort Dearborn Co.: Transforming from 'Most Wanted' to Best in Class." *Professional Safety*, March 2004: 25–31.

Scholtes, Peter R. and Russell L. Ackoff. *The Leader's Handbook: Making Things Happen, Getting Things Done.* New York: McGraw-Hill, 1997.

Secretary of Labor v. E Smalis Painting Co., Inc. Occupational Safety and Health Review Comission, Docket No. 94-1979. April 10, 2009. http://www.oshrc.gov/decisions/pdf_2009/94-1979.pdf

Torres, Katherine. "Making a Safety Committee Work for You." *Occupational Hazards*, October 23, 2006.

U.S. Department of Labor. "The Occupational Safety and Health Act of 1970." *General Industry Standards*, vol. 6, chap. XVII, 29 CFR pt. 1910. Washington, D.C.

U.S. Department of Labor. Bureau of Labor Statistics. "Injuries, Illness, and Fatalities." Accessed April 13, 2015. http://www.bls.gov/iif/

U.S. Department of Labor. Bureau of Labor Statistics. "2010 Injury and Illness Statistics." January 2012. Accessed April 30, 2015. http://www.printing.org/print/10121

U.S. Department of Labor. Occupational Safety and Health Administration. "OSHA Inspections." OSHA 2098, 2002 (revised). Accessed April 13, 2015. https://www.osha.gov/Publications/osha2098.pdf

Weibert, Michael and Catherine Plunkett. "Motivational Safety System: One Organization's Experience Moving Toward World-class Performance." *Professional Safety*, February 2006: 34–39.

Resources

Drebinger, John. *Would You Watch Out for My Safety?*
https://www.drebinger.com/shopping-cart/would-you-watch-out-for-my-safety/

E5 Incidence Rates
https://www.rit.edu/~w-outrea/OSHA/documents/Module5/M5_IncidentRates.pdf

Fort Dearborn Company Leadership in Safety Award
http://www.fortdearborn.com/news-events/

Gallup Q^{12} Employee Engagement Survey
https://q12.gallup.com/Public/en-us/Features

Acronyms

AED	Automated External Defibrillator
ASSE	American Society of Safety Engineers
BBS	Behavior-based Safety
CEO	Chief Executive Officer
CFO	Chief Financial Officer
COI	Certificate of Insurance
COO	Chief Operating Officer
CPR	Cardiopulmonary Resuscitation
DART	Days Away Restricted Transferred
DNA	Deoxyribonucleic Acid
EBITDA	Earnings Before Interest, Taxes, Depreciation and Amortization
EHS	Environmental Health and Safety
EPA	Environmental Safety Administration
FDC	Fort Dearborn Company
GM	General Manager
HR	Human Resources
IH	Industrial Hygiene
IR	Incident Rate
JSA	Job Safety Analysis
KPI	Key Performance Indicators
LCL	Lower Control Limit
LLC	Limited Liability Company

LOTO	Lock Out Tag Out
LTC	Loss Time Case
LWDI	Loss Workday Incidence Rates
LWDII	Lost Workday Injury and Illness Rate
MBA	Master in Business Administration
MOD	Experience Modifier
MS	Master of Science
NCCI	National Council of Compensation Insurance
NFPA	National Fire Protection Agency
OSHA	Occupational Safety and Health Administration
OTIF	On Time in Full
POTW	Publicly Owned Treatment Works
PQ	Positive Quotient
PVIF	Present Value Interest Factor
QC	Quality Control
RCRA	Resource Conservation and Recovery Act
ROI	Return on Investment
SH&E	Safety, Health & Environment
SPY	Safety Professional of the Year
TTD	Temporary Total Disability
UCL	Upper Control Limit
VOC	Volatile Organic Compounds
VPP	Voluntary Protection Program
WCSC	World-class Safety Culture

Appendix

World-class Safety Culture: Applying the Five Pillars of Safety

When OSH professionals talk about world class, they generally mean best of the best, best in the class, best in the world. Murali (2012) defines safety culture as "the attitude, beliefs, perceptions and values that employees share in relation to safety" in an organization.

Since safety is a process, world-class safety cannot have a singular value. The Campbell Institute identified five main qualities based on its analysis of applications for NSC's Award of Excellence, which recognizes superior OSH management systems. Specifically, these are: 1) OSH on par with business performance; 2) system-based approach to OSH; 3) continuous improvement; 4) OSH aligned with organization strategies and values; and 5) promoting safety and health on and off the job.

Similarly, Hansell (2012) identifies five key qualities found among world-class companies: 1) visible senior management leadership and commitment; 2) employee involvement and ownership; 3) systemic integration of OSH and business functions; 4) data-based decision making and system-based root-cause analysis; and 5) going beyond compliance.

This article shares lessons from the author's experience in

helping a large multi-location printing operation develop a world-class safety culture (Saujani & Adler, 2004). This effort was based on five key pillars: 1) management commitment; 2) employee ownership; 3) system data; 4) system integration; and 5) organization-wide engagement. Although this case example involved a manufacturing setting, most of the principles and ideas used can apply to various other industries as well.

Management Leadership & Commitment

Visible senior management leadership and commitment to safety are critical factors in setting a goal to attain world-class performance and developing the culture needed to achieve this goal. This commitment is best indicated "by the proportion of resources (time, money, people) and support allocated to health and safety management and by the status given to health and safety" (Flin & Yule, 2004).

How can an OSH professional secure management commitment? A proactive safety professional should know the characteristics of senior managers and understand what distinguishes the organization from similar companies. Some senior leaders are holistic and may need constant communication. Others make knowledge-based decisions—they need to hear logical reasoning behind safety-related activities and expenditures.

However, all senior leaders want their organizations to succeed financially and to perform optimally. This raises the question: How can an OSH professional use what motivates senior leaders (e.g., profit) to gain visible management commitment in safety?

Table 1

OSHA Penalty Structure

Violation type	Penalty
Willful A violation that the employer intentionally and knowingly commits or a violation that the employer commits with plain indifference to the law.	OSHA may propose penalties of up to $70,000 for each willful violation, with a minimum penalty of $5,000 for each willful violation.
Serious A violation where there is substantial probability that death or serious physical harm could result and that the employer knew, or should have known, of the hazard.	There is a mandatory penalty for serious violations, which may be up to $7,000.
Other than serious A violation that has a direct relationship to safety and health, but probably would not cause death or serious physical harm.	OSHA may propose a penalty of up to $7,000 for each other-than-serious violation.
Repeated A violation that is the same or similar to a previous violation.	OSHA may propose penalties of up to $70,000 for each repeated violation.

Note. Adapted from OSHA Training Institute presentation.

Financial Benefits

OSH professionals must strive to communicate to managers the direct, positive correlation between investment in safety and subsequent return on investment (Weibert & Plunkett, 2006). Avoiding legal issues and their associated costs is also a motivating factor. The expectations of various governing bodies continue to increase, as do fines and criminal investigations for noncompliance. Table 1 shows the potential penalties imposed by OSHA for regulatory violations; these serve as a strong message of the benefits of a proactive safety culture.

Safety improvements can also increase productivity. For

example, during his work as a loss-control consultant, the author worked with a glove manufacturer that experienced one finger amputation every year for several years. These injuries occurred in an operation in which employees cut cloth with foot-operated clicker presses. Initially, the plant's general manager viewed these injuries as a cost of doing business. However, the insurance company (for which the author was a loss-control consultant) recommended that the company install two-hand trip devices. The company did so and subsequently reported that production increased more than 20 percent. So, in this case, eliminating a safety concern (amputation potential) minimized operational risks and delivered a financial benefit in the form of higher productivity.

Allocating the Cost of Injuries

Another technique available to OSH professionals is to allocate the cost of injuries to each plant or profit center. Different systems can be used to achieve this fairly and equitably. Most companies charge back workers' compensation costs to a plant based on its payroll; however, this approach is not sensitive to controlling injuries and associated costs. Other companies pay the workers' compensation premiums up front, then collect monies from each plant after the end of the policy year based on the percent cost of injuries per plant per year. This approach is an improvement since it encourages a proactive approach to controlling employee injuries and costs.

A better system is to collect workers' compensation insurance costs from each plant using a formula, then provide a rebate after the close of a policy year based on each plant's injury experience. The formula could have two components: fixed cost and variable cost. The fixed cost element could be the company's deductible premium for the large loss deductible program divided by the total company payroll multiplied by the plant's payroll. The variable cost could be the weighted cost of injuries measured by the incurred cost for the last 4 years. The variable cost of injuries could then be allocated based on a 4-year average allocation percent developed per plant. Table 2 presents an example of variable cost allocation. This approach requires active participation in managing safety by all general managers and employees.

Table 2

Plant Allocation

	Total incurred policy year 1	Total incurred policy year 2	Total incurred policy year 3	Total incurred policy year 4	Weighted average loss	Loss allocation %
Annual weight	10%	10%	40%	40%		
Plant A	395,395	242,288	114,687	35,305	123,765	32.23
Plant B	135,835	320,610	25,162	63,741	81,206	21.15
Plant C	19,280	2,668	0	72,691	31,271	8.14
Plant D	27,115	100,933	215,118	19,963	106,837	27.82
Plant E	543	282,441	1,861	0	29,043	7.56
Plant F	96,319	3,529	1,103	3,578	11,857	3.09
Gross allocable cost	674,487	952,469	357,931	195,278	383,979	100

Return on Investment: Cost-Benefit Analysis

Another way to secure senior managers' support for a new safety project is to provide them with ROI details that highlight how the company would benefit from investing in that project. ROI is the financial measure commonly used to compare investment opportunities. Most companies require ROI calculations for all investments, and such calculations must pass a hurdle rate, or minimum rate of return (e.g., 8 percent to 9 percent) to be acceptable (Pais, 2011).

For example, the finishing department in a printing company was reporting ergonomic injuries. Employees in this department used five paper-cutting machines equipped with two-hand trip devices to cut paper. Once the cutting cycle was complete, employees would pick up waste paper trims with their right hand and throw the trimmed paper into a cardboard box placed behind them. Once the box was full, a lift truck operator would dump its contents in a baler; the bales of paper trims were later picked up by a waste hauler for recycling.

This process required paper cutters to assume awkward postures and use awkward shoulder movement.

Helpers had to bend down to pick up the waste trim, causing some back injuries. The safety manager analyzed the process and recommended installation of an automatic vacuum system that would pick up the trims from all five finishing cutters and move them to the baler via air ducts. However, because this system cost $250,000, the safety manager had to complete an ROI calculation to demonstrate the potential financial benefit to the company.

For most safety projects, ROI = risk reduction divided by cost. However, this project presented significant benefits and risk reductions, so this equation was used:

- ROI = [(benefits/time) + (risk reduction/time)]/the initial investment
- ROI% = (return – cost of investment)/(cost of investment) × 100

Table 3 presents an example ROI calculation from the author's experience.

Table 3
Return on Investment

Year	Item	Benefits	Cost reduction	Total savings
1	Helper not needed ($12.28 per hour plus benefits at 23% of rate, two shifts)	---	$60,418	$60,418
2	Increase cutter efficiency, 4%	$12,084	---	$12,084
3	Reduced waste	$1,200	---	$1,200
4	Savings in leasing lift truck	---	$2,178	$2,178
5	Reduced cost of ergonomic injuries	---	$25,360	$25,360
6	Good night's sleep, increase level of safety confidence	$5,000	---	$5,000
	Total			$106,240

Savings to Be Realized per Year

The term *good night's sleep* (Rodda & Hedges, 1983) is the value assigned for the reduced anxiety and added peace of mind achieved when one has identified and controlled loss exposures through safety improvements. This value is higher if the potential for a loss is high without the safety intervention. It is a subjective number and varies depending on the exposure, reduction in potential for a loss, and number of senior managers and employees affected.

Consider this example from the printing organization. For a 3-year period, based on insurance company loss analysis, savings of $560,387 ($1,259,345 − $698,967) was achieved throughout the six plants, with a projected savings of approximately $31,132 for this plant and an estimated savings of $25,360 (81 percent) in the finishing department (Figure 1). The result was calculated by dividing the ROI by the cost of the investment expressed as a percentage: 106,240/250,000 = 42 percent. In other words, 42 percent of the investment would be recovered in the first year; it would take about 28 months to recoup the total investment.

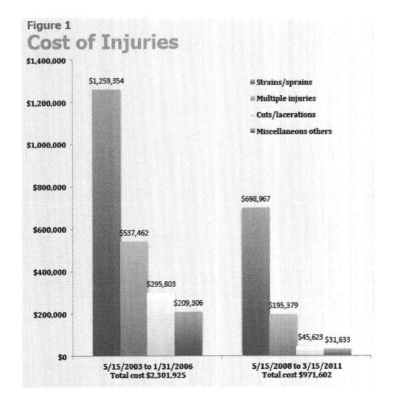

Figure 1

Cost of Injuries

Legend: Strains/sprains, Multiple injuries, Cuts/lacerations, Miscellaneous others

5/15/2003 to 1/31/2006
Total cost $2,301,925
- $1,259,354
- $537,462
- $295,803
- $209,306

5/15/2008 to 3/15/2011
Total cost $971,602
- $698,967
- $195,379
- $45,623
- $31,633

The 3-year ROI percent is often called the internal rate of return; it is calculated as net return (3 years' total returns minus the cost of investment) divided by the cost of investment expressed as a percentage. In this example, this was calculated at 27 percent [($106,240 × 3 − $250,000)/$250,000], or annual rate of return of 9 percent. This surpassed the company's 8.5 percent hurdle rate.

Various factors can influence ROI calculations. For example:

- Companies with a high cost of capital may have a higher hurdle rate.
- The years needed to recoup the investment may vary depending on the type of investment and the company's tax amortization policy.
- Present value factors must be considered if the net present value is included in the calculations.

Employee Involvement & Ownership

A company that is world class or is striving to be world class involves the entire organization in safety. This usually occurs through a series of initiatives. For example:

- Define roles and responsibilities for various levels in the organization (Saujani & Adler, 2004).
- Perform safety awareness surveys.
- Encourage and reward safe behaviors.
- Engage associates [e.g., safety committee meetings (Torres, 2006); risk assessments and safety audits; safety board program; identify-the-hazard program; safe plant of the year award; safety conversations].

Defining roles and responsibilities for each stakeholder helps to ensure that all personal are focused on a singular path of achieving safety success. For example, the role and responsibilities of a safety director could be: plan and implement company safety policy; coordinate company-wide safety initiatives; audit company facilities and operations; investigate incidents; support and encourage safety team leaders; train and motivate associates; and be a safety champion at the executive level.

Associates at all levels can be actively engaged through monthly safety committee meetings that follow a meaningful agenda. Asking committee members to volunteer for semimonthly plant safety audits adds another layer of involvement.

Managers can become engaged by completing daily informal inspections to identify and correct unsafe conditions and address risky behaviors in their departments. The safety director and general manager might be asked to conduct annual risk assessment surveys and establish safety goals for continuous improvement.

At the multiplant printing company, the safety board program helped encourage employee ownership and involvement. Under this program, employees in each of the organization's four departments had to prepare a safety board each quarter centered on a particular safety theme (e.g., electrical safety, HazCom, machine guarding, lift truck operator safety). The completed safety boards were then judged by a team of three senior executives. Employees in the department with a winning safety board received a free lunch.

The program generated extensive chatter and safety discussions among employees. The safety board program also led to a safe plant of the year award program. The goal was to encourage plant management to take an active interest in safety. Each plant had to submit documentation highlighting its initiatives and successes. The award selection committee, which consisted of the senior vice president of operations, corporate safety director and corporate human resource director, used the following criteria to score each plant's submission:

- results and accomplishments (30 points);
- general manager's commitment to safety (15 points);
- department managers' involvement in safety (15 points);
- fully functional safety committees (10 points);
- training and motivation of associates and managers (25 points);
- miscellaneous (5 points).

Integration of OSH & Business Functions

The systemic integration of OSH and business is reflected in several ways. For example, when the safety function has a direct reporting relationship with the CEO or COO and is involved in business decisions, it signals that safety is a critical business function. For example, when the firm was in the process of acquiring a new plant in a coastal area with significant exposure to hurricanes, the safety director (the author) recognized that the building was not designed for this exposure; he recommended that as a purchase condition the building's wind protection be significantly improved.

In another situation, the company purchased a large printing press with the intent of moving the press to one of its plants in 3 months. However, during this 3-month period, the seller was to continue to operate and maintain the press. The safety director intervened and required the seller to provide a COI naming the company as an additional insured while the press was in the seller's care and custody. A week after the closing a serious incident involving the press occurred; the safety director's actions saved the company significant money and spared its executives some tough questions from the board of directors. These examples illustrate how safety goals/strategies can align with company goals/strategies and create momentum toward world-class performance.

Data-Based Decision Making

A company must gather data so it can analyze the safety system and ensure that safety decisions are sound. Common data points include trends in incident rates (e.g., DART rates, lost-time incident rates); loss analysis trends; gap analysis for established safety goals; safety awareness scores; and hazard surveys (Middlesworth, 2013).

Perception surveys are another data source. These surveys help a company check how safety culture is perceived at the plant floor level, which can reveal gaps and identify areas that need improvement.

Two lagging indicators of note include the NCCI mod chart and OSHA incident rates. The NCCI mod chart shows a company's overall loss experience for a 3-year period and provides some indication of how well the company is managing incidents and injury costs. OSHA rates reflect how many incidents have occurred in that category in a year for an average of 100 employees. Such statistics primarily reflect past performance, but they can help a company compare how well its safety system is performing compared to its peers and to prior years.

To achieve world-class safety performance, companies must also identify and monitor leading indicators. Examples of these include completed behavior-based safety observations; completed safety training (managers and associates); completed safety audits and hazard assessments; corrected hazards; completed department safety meetings; and completed coaching and counseling sessions.

Figure 2 presents a road map that companies can follow to improve safety performance. By definition, truly world-class companies exceed compliance. They are proactive in identifying hazards, and thoroughly investigate incidents and near-hits to continuously improve the safety system.

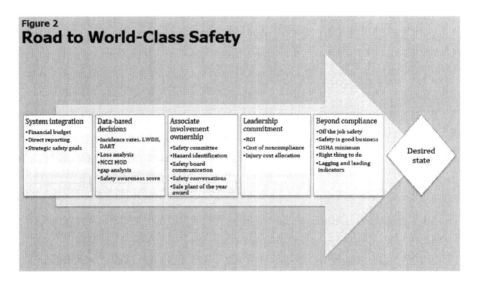

Figure 2
Road to World-Class Safety

System integration	Data-based decisions	Associate involvement ownership	Leadership commitment	Beyond compliance	
•Financial budget	•Incidence rates, LWDII, DART	•Safety committee	•ROI	•Off the job safety	Desired state
•Direct reporting	•Loss analysis	•Hazard identification	•Cost of noncompliance	•Safety is good business	
•Strategic safety goals	•NCCI MOD	•Safety board communication	•Injury cost allocation	•OSHA minimum	
	•gap analysis	•Safety conversations		•Right thing to do	
	•Safety awareness score	•Safe plant of the year award		•Lagging and leading indicators	

Michael Saujani, CSP, CPCU, retired in 2013 as corporate safety director for Fort Dearborn Co., and formed his own consulting firm, MKS Safety LLC. His loss-control experience includes positions with Amerisure Mutual Insurance, Hanover Insurance, Hartford Insurance and Allstate Insurance. Saujani holds a B.S. in Mechanical

Engineering, and he is a professional member and past president of ASSE's Northeastern Illinois Chapter, which named him its Safety Professional of the Year in 2012.

References

Ayers, D. (2006, June). Mapping support for an EHS management system. Retrieved from http://ehstoday.com/safety/ehs_imp_38310

Cable, J. (2005). At 12 of America's safest companies, managers and employees share commitment and responsibility for safety. Retrieved from http://ehstoday.com/mag/ehs_imp_37819

Campbell Institute. (2012). Defining world-class EHS: An analysis of leading EHS management system practices of Robert W. Campbell Award winners. Retrieved from www.nsc.org/CampbellInstituteandAwardDocuments/WP-DefiningWorldClass-EHS.pdf

Flin, R. & Yule, S. (2004). Leadership for safety: Industrial experience. *Quality & Safety in Healthcare*, 13, ii45-ii51, doi:10.1136/qshc.2003.009555. Retrieved from http://qualitysafety.bmj.com/content/13/suppl_2/ii45.full

Hansell, C.A. (2012). Become a transformational safety leader. Presentation at America's Safety Conference, Sept. 11-12, Chicago, IL.

Hansell, C.A. (2010). Successfully aligning and integrating safety within the business. *Proceedings of ASSE's Safety 2010*. Retrieved from www.asse.org/assets/1/7/601.pdf

Middlesworth, M. (2013, April 24). A short guide to leading and lagging indicators of safety performance. Retrieved from http://ergo-plus.com/leading-lagging-indicators-safety-preformance

Murali, J. (2012, Jan. 31). Road map to world-class safety for L&T. Retrieved from www.slideshare.net/vinuvinu/roadmap-to-world-class-safety-forNCCI. ABCs of experience rating. Retrieved from www.ncci.com/documents/abc_Exp_Rating.pdf?

Pais, D.S. (2011). Showing EHS value through return on investment. *Proceedings of ASSE's Safety 2011.*

Rodda, W.H. & Hedges, B.A. (1983). *Commercial property management and insurance.* Malvern, PA: American Institute for Property and Liability Underwriters.

Saujani, M. & Adler, N. (2004, March). Safety at Fort Dearborn Co.: Transforming from most wanted to best in class. *Professional Safety, 49*(3), 25-31.

Scholtes, P.R. & Ackoff, R.L. (1997). *The leader's handbook: Making things happen, getting things done.* New York, NY: McGraw-Hill.

Torres, K. (2006, Oct. 23). Making a safety committee work for you. Retrieved from http://ehstoday.com/safety/best-practices/ehs_imp_39324

Weibert, M. & Plunkett, C. (2006, Feb.). Motivational safety system: One organization's experience moving toward world-class performance. *Professional Safety, 51*(2), 34-39.

Senior Management's Commitment to Safety: A Critical Factor in Achieving Safety Results

BY MICHAEL SAUJANI, CPCU, CSP

Senior management's commitment to safety is one of the most important factors in achieving a safe and healthy work environment. The million-dollar question is: how can you secure and gauge senior management's commitment to safety?

Senior management is an individual or a team that is held responsible for effective performance of the operations of the company. Senior management's commitment to safety in a privately held small to medium-size corporation can be obtained by pointing out the financial benefits of the program, as the prime driver for these companies is the ability to make a profit. For a public corporation, although profit is a motivator, other factors, such as social responsibility, play a significant role in senior management's decision-making process.

In my analysis, companies use four techniques to secure senior's management commitment, thereby reducing the cost of doing business, increasing profits, and keeping OSHA at bay:

- Know their audience
- Recognize that safety makes good business sense
- Consider the cost of noncompliance
- Develop and appoint a meaningful safety system champion as a voice of safety for the company

A proactive safety professional should know the characteristics of senior management and what distinguishes its organization from others in the same arena. All leaders want their organizations to succeed financially and to be the best they can be. So as safety professionals, we need to make sure there is a financial benefit to the organization from our safety initiatives.

Manufacturing Operations

The United States is the world's largest economy, employing 140 million people in retail, energy, agricultural, manufacturing, and finance sectors.[13]

Since 1970, the federal government has been regulating private enterprises in order to achieve social goals such as improving the public health and safety and maintaining a healthy environment. OSHA[14] is responsible for improving public health and safety. It does this by promulgating and enforcing standards for workplace safety.

Every year, OSHA sends letters to its "most wanted" companies: those with injury rates that are higher than the industry average. In its letters, OSHA urges management "to remove hazards causing high rates."

Although many companies have regular safety meetings and

---

13 Wikipedia, "Economy of the United States," en.wikipedia.org/wiki/ Economy_of_the_United_States (accessed April 13, 2015).
14 United States Department of Labor (Occupational Safety and Health Act of 1970), General Industry Standards, vol. 6, chap. XVII, 29 CFR pt. 1910, Washington, D.C.

complete OSHA-required safety training, they may still find their cost of injuries going up. I believe the missing link is the commitment, preferably visible, to safety from senior executives and managers.

Financial Benefits

As safety professionals, we are responsible for alerting managers to the direct, positive correlation between investment in safety and return on investment.

There is evidence that companies that implement effective safety programs can reduce injury and illness rates by more than 20 percent and generate a return of more than four dollars for every one dollar invested in safety. These companies are also able to avoid OSHA penalties in case of an audit. Safety is a non-delegable duty; per OSHA directive, senior executives and general managers cannot delegate safety responsibilities to line supervisors.

The expectations for private organizations by various governing bodies in the U.S. and several other countries have increased significantly. Safety regulations have become more complex with criminal investigations for noncompliance and significant OSHA fines. The "Citations and Penalties" chart offers a look at OSHA's penalty structure.[15]

Citations and Penalties

Violation Type	Penalty
WILLFUL A violation that the employer intentionally and knowingly commits or a violation that the employer commits with plain indifference to the law.	OSHA may propose penalties of up to $70,000 for each willful violation, with a minimum penalty of $5,000 for each willful violation.

15 United States Department of Labor, Occupational Safety and Health Administration, "OSHA Inspections, OSHA 2098," OSHA, 2002 (revised) www.osha.gov/Publications/osha2098.pdf (accessed April 13, 2015).

Violation Type	Penalty
SERIOUS A violation where there is substantial probability that death or serious physical harm could result and that the employer knew, or should have known, of the hazard.	There is mandatory penalty for serious violations which may be up to $7,000.
OTHER-THAN SERIOUS A violation that has a direct relationship to safety and health, but probably would not cause death or serious physical harm.	OSHA may propose a penalty of up to $7,000 for each other-than-serious violation.
REPEATED A violation that is the same or similar to a previous violation.	OSHA may propose penalties of up to $70,000 for each repeated violation.

With professional safety guidance from in-house or private safety consultants, the company's bottom line may be significantly strengthened by compliance to safety standards.

The cost of a willful violation—a violation that the employer allegedly intentionally and knowingly committed—may each cost up to $70,000. The company and responsible management individuals, including general and department managers, if convicted in criminal proceedings of a willful violation, can be fined up to $500,000 and receive penalties of up to six months in jail.

Improvements in safety improve productivity and mitigate fines because employees who feel safe often work harder. For example, when working for an insurance company as a loss-control consultant, I remember a glove manufacturer that experienced one finger amputation every year for several years while employees were cutting cloth with foot-operated clicker presses. For a while, to the general manager at that time, the losses were simply a cost of doing business. Then the insurance company recommended that the company implement two hand-trip devices.

Once the devices were implemented, production went up by more than 20 percent, as employees no longer had to worry about potential finger amputations. So the improvements in safety also improved financial performance and minimized operational risks for the company. Ensuring the safety of employees (and customers) while on company premises is not only the company's moral and legal obligation, but it also makes good business sense.

Several costs to the company can result from unsatisfactory safety precautions:

- Insurance costs – increased workers' compensation premiums
- Cost of employees not working – hiring and training temps
- Lost production and overall increased cost of doing business
- Potential quality issues
- Social responsibility – being a good employer, community asset, desirable place to work

Senior management's commitment can be secured for new safety projects if they receive calculations showing the ROI and how the company would benefit from investing in the safety project. The ROI is the financial measure[16] commonly used to evaluate attractiveness of one investment over another. Most companies require ROI calculations for all investments, and such calculations should pass a hurdle rate, or minimum rate of return (often 8 to 9 percent), to be acceptable.

For example, a printing company was having significant ergonomic injuries in its Finishing Department. The Finishing Department had five paper-cutting machines that required employees to cut paper using two hand-held devices. Once the cutting cycle was complete, employees would pick up the waste paper trims with their right hand and throw the trimmed paper in a cardboard box placed behind them.

Later, when they had time, they or a helper would pick up the waste trim from the floor and place it inside the cardboard box. Once the box was full, a lift truck operator would dump it in a baler.

16 Anthony Veltri and Jim Ramsay, "Economic Analysis: Make the Business Case for SH&E," *Professional Safety*, September 2009, pp. 22–30.

The bales of paper trims would be picked up by a waste hauler for recycling.

This process required awkward posture and shoulder movement by the paper cutters, often causing severe back and shoulder injuries. It also required the helpers to bend down and pick up the waste trim, causing some back injuries. In addition, the lift truck operator could potentially hurt the finishing operators while picking up the boxes of paper trims.

The safety manager analyzed the process and recommended an automatic vacuum trim system that would pick up the trims from all five finishing cutters and move them to the baler via air ducts. However, the cost of the system was $250,000, and it was critical to show the financial benefit by completing the ROI calculation:

$$\text{ROI} = \text{Benefits/Time} \div \text{Initial investment}[17]$$

For most safety projects, ROI equals risk reduction divided by cost. However, for this project, there were significant benefits and risk reduction, so ROI looked like this:

$$\text{ROI} = (\text{Benefits/Time}) + (\text{Risk reduction/Time}) / \text{Initial investment}$$

The table "Savings Realized per Year" shows the benefits over time and the savings in risk reduction.

---

17 David S. Pais, "Showing EHS Value Through Return on Investment (ROI)," presentation at the ASSE Professional Development Conference and Exposition Safety 2011, June 15, 2011, Chicago.

Savings Realized per Year

Year	Item	Benefits	Cost Reduction	Total Savings
1	Helper not needed ($12.28 per hour plus benefits at 23% of rate—two shifts)		$60,418	$60,418
2	Increase cutter efficiency—4%	$12,084		$12,084
3	Reduced waste	$1,200		$1,200
4	Savings in leasing lift truck		$2,178	$2,178
5	Reduced cost of ergonomic injuries		$25,360	$25,360
6	Good night's sleep—increased level of safety confidence	$5,000		$5,000
	Total			$106,240

A "good night's sleep" in risk management is the value assigned for senior management to sleep well as a result of a safety-related improvement. This value is higher if the potential for a loss is high without the safety intervention. It is a subjective number and varies depending upon the exposure, reduction in potential for a loss, and senior management affected.

For the three-year period, based on insurance company loss analysis, a savings of $560,387 was realized throughout the six plants, with a projected savings of approximately $31,132 for this plant and an estimated savings of $25,360 in the Finishing Department.

The ROI for this project was calculated to be 106,240/250,000, or 42 percent every year. In other words, it would take about two years and four months to recoup the investment. The three-year rate of return was calculated at 127 percent ($106,240 × 3/$250,000), handily surpassing the annual company hurdle rate of 8.5 percent. If the annual hurdle rate (minimum acceptable rate of return) was higher than 9 percent, then this project would have been denied if a decision was made purely on a financial basis.

There are various factors that would change the ROI calculations, including:

- The hurdle rate could be higher for companies whose cost of capital is high.
- The years needed to recoup the investment may vary depending upon the type of investment and the company's tax amortization policy.
- The present value factors, if the net present value is included in the calculations.

Safety Systems

A sustained safety effort and management commitment is contingent on an effective safety system. One printing company put in place a ten-point safety and risk management system, dubbed SENIORMGMT. This system included:

S – Safety champion/risk manager

E – Engage through safety committee[18]

N – Necessary safety audits

I – Insurance and risk management program[19]

O – Open communications

R – Root cause analysis

M – Mission statement

G – Gauging or monitoring of the safety process[20]

---

18 Katherine Torres, "Making a Safety Committee Work for You," *Occupational Hazards*, October 23, 2006.

19 Josh Cable, "At 12 of America's Safest Companies, Managers and Employees Share Commitment and Responsibility for Safety," *Occupational Hazards*, September 21, 2005.

20 David A. Jones, "Mapping Support for an EHS Management System," *Occupational Hazards*, June 29, 2006.

M – Motivate senior management and employees[21]

T – Training and education for managers and associates[22]

Safety Champions

As safety professionals, we should strive to have safety champions at all levels of our organizations. The CEO should be the safety champion representing the company to the board of directors. The senior vice president of operations should be the safety champion representing the manufacturing operations and getting on board the plant general managers who report directly to him.

The safety director/manager should be the safety champion for the company for internal communication with executives and external communication with insurance companies and various regulators. The general managers and their safety team leaders should be safety champions at the plant level. The managers should be safety champions counseling and motivating the associates to be safe. The associate safety champions should participate in safety committee meetings, complete safety audits, and prepare safety boards for safety awareness campaigns. Every company needs these safety champions to spread the word for a safer and more productive work environment.

---

21 Michael Saujani and Nick Adler, "Safety at Fort Dearborn Co.: Transforming from 'Most Wanted' to Best in Class," *Professional Safety*, March 2004, pp. 25–31.

22 M. Weibertand Catherine Plunkett, "Motivational Safety System: One Organization's Experience Moving Toward World-class Performance," *Professional Safety*, February 2006, pp. 34–39.

Management Commitment/Employee Involvement

Increased, visible levels of management's commitment to safety should help to control the cost of injuries and make operations safer. Senior management's commitment to safety at the corporate and plant levels is the most important factor in achieving a safe and healthy work environment. It is difficult to measure this commitment, and there is no easy way to secure it, but once it is achieved, outstanding safety results do follow. In addition, employees are more inclined to follow safety guidelines when they know that upper management is supportive.

Corporate safety directors of proactive safety companies can show this commitment by being visibly involved in completing the annual safety audits and presenting their findings to their managers as well as the general managers. They can help develop and present OSHA required and other safety training programs and should be visibly present at audit presentations by regulators or insurance providers, taking the responsibility for any corrective actions.

Department managers can show their commitment by being visibly involved in behavior-based safety observations and by completing coaching or counseling sessions. They can participate in all accident investigations in and complete daily safety checks of their departments, making sure hazards are identified and eliminated.

Knowing that American workers are highly competitive and aiming to increase managers' commitment to safety, one printing company developed a program to offer the Safe Plant of the Year award. Each plant within the company was required to submit documentation showing its initiatives and successes. The Safe Plant of the Year selection committee consisted of the printing company's senior vice president of operations, corporate safety director, and corporate human resources director. The criteria for scoring the plants' accomplishments included:

- Results and accomplishments – thirty points
- General managers' commitment to safety – fifteen points
- Department managers' involvement in safety – fifteen points
- Fully functional safety committees – ten points
- Training and motivation of associates and managers – twenty-five points
- Miscellaneous – five points

A traveling trophy was presented to the winning plant by the chief executive officer. All associates at the plant who were selected to receive the award were honored with an event that included a catered lunch paid for by the company. The program helped the printing company increase the level of commitment from department managers and employees, creating an exciting yet safe atmosphere for work.

In addition, this visible management commitment for the company was publicly noted by its insurance company consultants. In his 2010 report, Account Executive Brett Carroll, of The Hartford Insurance Company, acknowledged the company as follows:

"... although there was a very strong commitment and formal program in place, the company was eager to reach out and learn in an effort to continually strengthen their program. Secondly, we quickly realized that this program had full support of management. A management influence is present in all safety activities, and there is no doubt that safety is a key component of their business."

Accomplishments

It is clear, then, that the visible management commitment to safety has significant benefits. The proof is in the pudding.

In the example provided, the printing company had proactive, visible senior management commitment. Its employee injury experience for the last five years is summarized in this loss analysis chart:

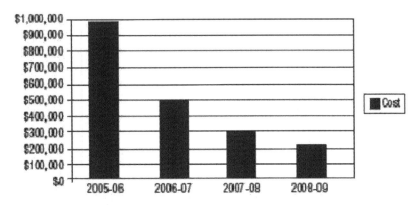

Company Loss Experience

To gain your own safety-related accomplishments, you can first invite safety consultants from states' outreach programs to identify hazards and recommend controls at your company. If, after taking this action, you had an unannounced visit from an OSHA compliance officer, you would be in a better position to manage that visit. In fact, at one company, as a result of developing and implementing a comprehensive safety and training program, when OSHA compliance officers made three unannounced visits to three of its plants over the last three years, no citations were issued.

Successful safety contribution generally leads to financial benefits for employees, such as increased bonuses or raises. But more importantly, employees go home safely to their families,[23] and companies can see yearly reductions in their workers' compensation experience modification factor, or mod.

Mod is a reflection of actual loss versus the expected loss for the

---

23 United States Department of Labor, Bureau of Labor Statistics, "Injuries, Illness, and fatalities," www.bls.gov/iif/ (accessed April 13, 2015).

same class of business. One company's mod was reduced from 1.29 to 0.75 over a four-year span, following senior management commitments. This reduction in mod reflects continuous improvement in safety results.

With senior management's[24] visible commitment to safety increased, workers' compensation injury costs and premiums would decrease. The NCCI MOD,[25] IR, and LWDI rates[26] would decrease as well. OSHA would no longer be on that company's doorstep every year.

---

24 Peter R. Scholtes and Russell L. Ackoff, *The Leader's Handbook: Making Things Happen, Getting Things Done* (New York: McGraw-Hill, 1997).

25 "ABCs of Experience Rating," NCCI, www.ncci.com/documents/abc_Exp_Rating.pdf (accessed June 2, 2011).

26 United States Department of Labor, Bureau of Labor Statistics, Printing Industries of America, 2010 Injury and Illness Statistics, www.printing.org/print/10121 (accessed April 13, 2015).

Safety at Fort Dearborn Company: Transforming from "Most Wanted" to Best in Class

BY NICK ADLER, EXECUTIVE VP AND
MICHAEL SAUJANI, CSP, CPCU, CORPORATE SAFETY
DIRECTOR

Fort Dearborn Company (FDC) is a privately held family owned corporation with multiple facilities and manufacturing systems throughout the world. We provide a full range of high quality labels for our customers that include the food, beverage, paint and retail industries. Our manufacturing systems have the capability to provide our customers with HiColour cut and stack labels, film labels, pressure sensitive labels and digital printed labels. Most of our (ergonomic and material handling related) accidents have occurred at our facilities that cut and stack labels.

The cut and stack labels are made from rolls of paper that are fed through special machines to make sheets of paper of the desired dimensions. Palletized sheets of paper are used to print the desired labels on any one of our printing presses. The printed labels are cut using cutting machines, and then stacked and wrapped with cellophane plastic material and shipped to customers via common carrier.

The shrink sleeve labels are made from rolls of plastic that are rolled into place on any one of flexographic presses. Each press has several printing stations. As plastic material is fed through each printing station, ink is applied to make the desired labels. The rolls of labels are slit, packaged and shipped to the customers.

We were on Occupational Safety and Health Administration's (OSHA) "most wanted" list of companies whose injury rates were higher than the industry average. At the beginning of year 2002, two of Fort Dearborn Company's divisions received letters from Assistant Secretary for Occupational Safety and Health stating: "the Agency used this (accident and injury) data to identify the workplace with the highest LWDII; your workplace was one of those identified. This means employees in your businesses are being injured at a higher rate than in the most other businesses in the country." Our LWDII rate at these two facilities referenced to in the correspondence were 6.61 and 7.63! Actually, almost all of our facilities had higher than average LWDI. We knew we had to do something fast.

Our worker's compensation insurance company was not happy with our loss records and decided to increase our rates significantly. To give you an idea of the magnitude of increase, in 1997-98 we paid a premium in access of $220,000. By the year 2001-02 our premium had risen to 1.2 million dollars a whopping 491 percent increase! Obviously, FDC had to be aggressive in decreasing the cost of insurance. The Executive Directors had a meeting with the insurance agency and as requested by the agency, the executive committee authorized the hiring of a professional safety director to manage and turn around the company's poor loss experience. That was a first for the company and some executives even questioned the need for the additional overhead of hiring a professional.

FDC Premium Development

- The premiums we paid for workers compensation insurance has steadily increased.
- The associates working for FDC have not increased by the same percentage.
- We need to address this by prudent insurance and safety programs.

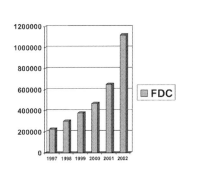

The general belief among the executives was that anyone with good common sense should be able to manage the company's safety program, not realizing that the company had grown to where it was no longer a small "ma and pa plant." The insurance agency executive emphasized that all of the other important departments such as Quality Control, Accounting, Finance, Marketing and Production were being managed by qualified experienced managers. Then why should the important function of safety and protecting company assets not be managed by a qualified manager? Within less than a year of hiring a professional safety director, with the accident rates tumbling, these executive realized "Safety Makes a Whole Lot of Business Sense."

Lake Forest - LWDII Annual Rate

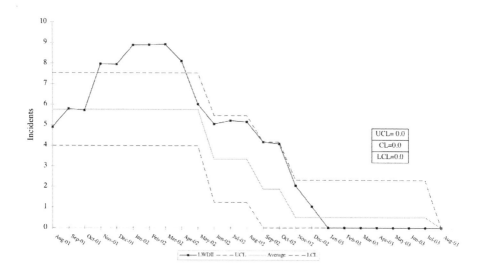

We would like to share with the safety and insurance professionals how we were able to reduce our insurance costs and incidence rates significantly so that their companies can reap the results of a safer plant as well.

The first and foremost thing that we did was to develop our "Current State" and "Desired State" and the Fishbone Diagram showing the systems we would want developed to achieve the "Desired State." Our Current and Desired State a year and half ago were:

Current State

- Insurance mod was 1.5; that is, we were paying 50 percent more than the average commercial printer.
- Worker's compensation premium had increased 425 percent or from $284,000 to $1.2 million in five years.
- Property premium had increased from $39,750 to $285,000 in five years.
- We did not have an effective, consistent Safety Policies and Procedures in place.

- We did not have a good Accident Investigation Program.
- We did not have monthly Safety Committee Meetings at all locations.
- We had volunteers serving as Safety Team Leaders at each facility.
- We did not have a professional Safety Director.
- Associates and managers were not fully recognized for their safety efforts.
- All department managers were not fully involved in safety.
- All division presidents were not fully involved in safety.
- Safety had not been a priority.
- Safety culture was inconsistent.
- Safety responsibilities were delegated to plant engineering.
- Hazards were not being corrected immediately.
- Associate training was inconsistent.
- Records of training and inspections were not available.

Desired State

- Safe and healthful working condition for our associates and preserve our human resources. This will be measured by FDC having an accident record one of the ten best in the commercial printing industry.
- Full management and associate participation in safety.
- Worker's compensation insurance cost to be no more than $720,000.
- Maintain worker's compensation mod to no more than .9 in three years.
- Zero serious safety violations and less than three non-serious safety violations at all facilities.
- Zero Loss Time accidents for the year and less than three No Lost Time incidents for every 100 employees.

Having determined our "desired state," we worked on developing safety systems to achieve this state. We tackled the problem

Fishbone Diagram: Corporate Safety Program Key Projects (10/21/02)

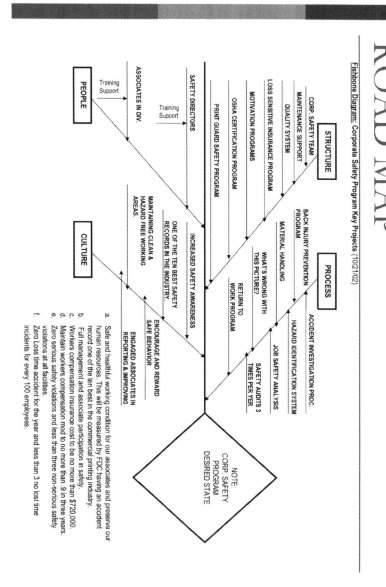

Fishbone Diagram: Corporate Safety Program Key Projects (10/21/02)

ROAD MAP

PEOPLE

Training Support

Training Support

ASSOCIATES IN DIV.

STRUCTURE

SAFETY DIRECTORS

CORP. SAFETY TEAM

MAINTENANCE SUPPORT

QUALITY SYSTEM

LOSS SENSITIVE INSURANCE PROGRAM

MOTIVATION PROGRAMS

OSHA CERTIFICATION PROGRAM

PRINT GUARD SAFETY PROGRAM

CULTURE

MAINTAINING CLEAN & HAZARD FREE WORKING AREAS.

ONE OF THE TEN BEST SAFETY RECORDS IN THE INDUSTRY.

INCREASED SAFETY AWARENESS

ENCOURAGE AND REWARD SAFE BEHAVIOR

ENGAGED ASSOCIATES IN REPORTING & IMPROVING

PROCESS

BACK INJURY PREVENTION PROGRAM

MATERIAL HANDLING

WHAT'S WRONG WITH THIS PICTURE?

RETURN TO WORK PROGRAM

ACCIDENT INVESTIGATION PROC.

HAZARD IDENTIFICATION SYSTEM

JOB SAFETY ANALYSIS

SAFETY AUDITS 3 TIMES PER YER

NOTE:
CORP. SAFETY
PROGRAM
DESIRED STATE

a. Safe and healthful working condition for our associates and preserve our human resources. This will be measured by FDC having an accident record one of the ten best in the commercial printing industry.
b. Full management and associate participation in safety.
c. Workers compensation insurance cost to be no more than $720,000.
d. Maintain workers compensation mod to no more than .9 in three years.
e. Zero serious safety violations and less than three non-serious safety violations at all facilities.
f. Zero Loss time accident for the year and less than 3 no lost time incidents for every 100 employees.

based on the structure we needed, the safety processes that needed to be implemented, the people we needed at each division and the type of a culture we were planning to develop.

We feel that safety is everyone's responsibility and wanted to include everyone, including shareholders, executives, managers, union and nonunion employees. As such, we developed a Progress Report and attached it to every payroll check defining the role of each associate.

As the chief operating officer of the corporation, the role and responsibility of the **FDC President** is to furnish each associate a place of employment which is free from recognized hazards that could cause injury of any type, serious or minor, to our associates and visitors.

The role and responsibility of the **Corporate Safety Director** who is in charge of overall company safety related goals and objectives are to:

a. Plan and implement company safety policies and procedures to comply with government rules and regulations.
b. Coordinate company-wide programs to ensure work site safety practices.
c. Audit FDC facilities to detect existing and potential accident and health hazards and offer advice to prevent these hazards.
d. Assist in the investigation of lost time accidents.
e. Analyze accidents and develop trends for the purpose of reporting corrective action.
f. Support safety team leaders to achieve company safety goals.
g. Coordinate the development of programs and procedures to ensure safe behavior.

In regard to safety, the role and responsibility of each **Division President or General Manager** are to support, encourage and lead the safety team leader, department managers and associates in fulfilling the Safety AIM or the desired state of the company.

The role of the division **Safety Team Leader**, who also has other management responsibilities, is to:

a. Complete required safety training of associates and maintain records of the training
b. Chair the safety committee meetings
c. Complete semi-monthly safety inspections using safety checklist
d. Investigate all accidents and complete a report of the investigation with corrective action
e. Motivate associates to work safely.

The role of each **Department Manager** in the division is to:

a. Complete weekly tool box meetings.
b. Complete a daily safety inspection of the department
c. Make suggestions on how to improve the system.
d. Complete daily safety huddle.
e. Assist in the investigation of accidents by Safety Team Leader and Safety Director.
f. Train associates in safe work practices.
g. Observe associates working safely.
h. Coach and counsel associates to ensure safe behavior.

The role of each **Associate** (employee) is to ensure a safe work place by working together with managers and division safety team leaders. Their role is to:

a. Obey all safety rules and regulations.
b. Participate in the training meetings.
c. Make suggestions on how to improve the system.
d. Work safely as an individual and as a team.
e. Ensure work area is kept clean and free of hazards.
f. Report all unsafe acts and conditions to the immediate supervisor.

Nick and I discussed the role and responsibilities at various meetings held at corporate and divisions so that the senior executives, managers and associates were aware of what was expected of them. At safety team leader's meetings, we discussed the importance of developing division safety committees and ensuring the committee members complete all desired safety activities. At the manager's meetings we discussed the importance of making sure all managers are trained in root cause analysis, accident investigations and 10 Hour OSHA certification program. When meeting with the associates, we made sure they understood the safety rules and discussed how the safety observation program will be used to comply with the rules. We provided gift certificates and safety products as awards to associates for safety observations and for hazard prevention ideas. As of this year, the General Manager's bonus (10 percent) will be determined by safety related accomplishments – reduction in LWDI and reduction in Incurred Cost of accidents versus the Expected Average Cost.

We met with the union leaders and got them to buy into our safety initiatives. We met with the executives and changed how we charged each division with the Worker's Compensation insurance cost. Previously each division paid to FDC corporate a contribution based on workers' compensation payroll. As a result, the division presidents (now, general managers) did not have any real incentive to manage the cost of worker's compensation insurance. We changed that. We purchased a Large Deductible insurance program, requiring us to pay a fee for insurance service and pay Workers Compensation cost incurred by the insurance company within 30 days of payment. We made each division responsible for its own Workers Compensation costs. At the end of the year, each division was charged back a percentage of the service fee based on payroll and the total cost incurred for all injuries at that division for the policy year. As a result, the division presidents watched for all claims and made sure conditions or processes causing the injuries were properly managed. At each of our facilities, a manager instead of a volunteer associate was designated as a Safety Team Leader to manage the safety program.

One of the most important responsibilities of the Safety Team leaders is to complete trending of employee injuries for each of the divisions they represented. They completed a trend report every month trending injury IR and LWDII. We initially used the Process Control Chart Tool Kit (PCCTK) developed by Sof-Ware Tools and later this year started using The QI Macros for Excel to monitor the safety process. The PCCTK and QI Macros were already being used by Quality Control and Production departments to track various operations and processes throughout the organization. So as to monitor the safety process, we used the XbarR – Average and Range chart – and XmR – the individual and Moving Range charts – (49, PCCTK.14-17 QI Macros) to evaluate the process stability. The XbarR chart provided the average rates and the Upper and Lower Control limits while the XmR chart provided process stability information. These charts helped us analyze if there were any "Special Causes" (28, Sholtes) influencing the safety process so we could jump on the situation as soon as possible. There are eight tests (147, PCCTK) of which the four listed below are most important to analyze if the process is stable and predictable. If the data on the chart violate any of the following four basic tests, the process is unstable and should be evaluated for special causes:

a. 1-point above or below the UCL or LCL
b. 8 consecutive points grouped above or below the average
c. 6 consecutive points ascending or descending
d. 14 consecutive points alternating up and down

Shown below is a LWDI chart from our largest division. We got similar data from all of our divisions. We then developed a combined chart for the corporation to evaluate how the safety system was performing at each division and companywide. We found out what gets measured gets done! The company focus on safety was paying off.

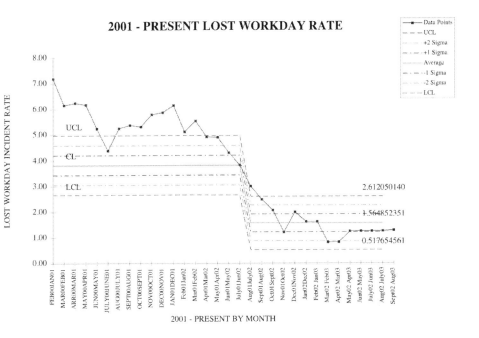

Our injury IR and LWDI started dropping since we started the program. Currently, as of August, 2003, the company wide LWDI is 0.48 while industry average is 2.7 for 200,000 man-hours worked. Our cost of insurance dropped from 1.2 million dollars to $619,076 for the first year and for the current four months in the policy the cost is $83,515! (Commercial Loss Experience, August 31, 2003).

Three of our divisions worked more than 200,000 man-hours without a lost time accident and received safety awards from Amerisure Insurance Company. One of our divisions worked more than six years without a lost time accident. We are proud of associates at these divisions who took the leadership in showing that if we work diligently, as a team, we can achieve outstanding safety results.

The primary reason we were able to make such significant progress was because we have developed safety policies that are implemented uniformly throughout the corporation.

- All of our divisions require monthly safety meetings that include associates from all departments. The members review accidents reports and how these accidents will be prevented in future. They review the results of the monthly safety inspections and actions plant engineering has taken to prevent hazards and also monitor progress in regards to other safety related activities. The minutes of these safety meetings are being developed and posted for all associates to read.

- Selected members of the Safety Committee (now) complete monthly safety inspections using the PrintGuard Safety Self-Inspection Checklist (A-1, PrintGuard Comprehensive Safety and Health Program Manual) to identify unsafe acts and conditions for corrective action by the plant engineering departments.
- The managers have completed daily informal inspections of their departments to identify and correct unsafe conditions. Each department manager was provided with a checklist (4.1 Informal Self Inspections. FDC Safety Manual) and then initial on the day of the month when the inspection was completed.
- We invited Illinois OSHA inspectors during the months of April and May of 2003 to inspect our operations for hazards, so we could take safety to the next level. One of the inspectors had about ten serious and more than seventy other than serious recommendations. All of these recommendations were corrected and a response sent requesting a revisit to verify appropriateness of our actions.

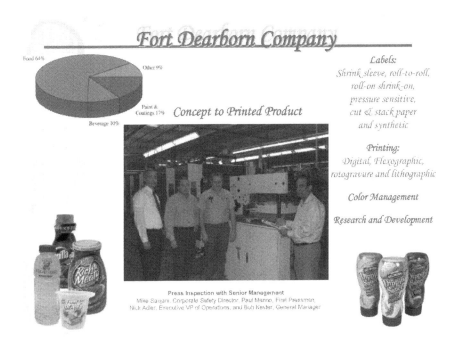

Fort Dearborn Company

Food 64%
Other 9%
Paint & Coatings 17%
Beverage 10%

Concept to Printed Product

Labels:
Shrink sleeve, roll-to-roll,
roll-on shrink-on,
pressure sensitive,
cut & stack paper
and synthetic

Printing:
Digital, Flexographic,
rotogravure and lithographic

Color Management

Research and Development

Press Inspection with Senior Management
Mike Saugani, Corporate Safety Director, Paul Manno, First Pressman,
Nick Adler, Executive VP of Operations, and Bob Kester, General Manager

- We have visited all of our divisions at least three times within the last year for the purpose of inspecting the plant operations for hazards and making safety presentations (Loss Analysis, Accident Investigations, Return to Work Program, etc.) at the managers' meetings. The reports of the surveys with recommendations were submitted within a week of the survey to the General Managers for corrective actions. These recommendations have been promptly complied with.

- Three times last year, we invited safety team leaders and human resources representatives for one-day seminars on safety. During these meetings, we discussed Ergonomics, Hazard Communication, Lock Out Tag Out, Accident Investigation, Property Conservation, and Fire Safety and Evacuation.

- All of our managers carry a "10 Hour OSHA Training" certificate so they understand the importance of complying with OSHA standards and identifying and correcting hazards.

- We have implemented "Return to Work" program to care for the injured associate and make every effort to bring the associate back to meaningful employment. When a first pressman strained his shoulder, we interviewed him and found he had good computer skills. We brought him back to work training our associates in Globe-Tek computer program we were installing to monitor our processes. We teach our managers to think about an injured person's ability and not his disability while attempting to bring him/her back to work.

- Safety is now being discussed at managers' meetings, the plant-wide meetings and employee discussions. All of our General Managers and department managers are involved in safety activities. When the general managers have their monthly plant wide meetings, monthly safety accomplishments and issues are the first being discussed. The department managers discuss safety along with other production and quality issues at their monthly department meetings. We have overheard associates discuss safety topics during lunch breaks.

- One of our divisions installed an automatic lifting device called VacuMove to lift boxes in the palletizing area using vacuum rather than associates' backs. One of our associates, Reynaldo Aragon said, "Now I am very happy, satisfied, and I don't worry because the VacuMove is fast, reliable, and safe to use." "I would like to thank Fort Dearborn's management for the idea to use this device in our plant. "

- We offered small rewards such as safety mugs, key chains, etc., for job observations and associate safety participation.

- Twice a year we survey our associates to evaluate how engaged they are in the success of our operation. In addition, once a year the corporate safety department completes a survey of how the associates are engaged in safety. We devised a checklist of 25 safety related questions and provided a copy of the survey with envelops at random to about 12 percent of the total division population. The associates responses were mailed to corporate safety department where these were tabulated and scored on a scale of one to five. The following were the results of the last survey:

Fountain Inn – 4.3
Niles – 4.1
Lake Forest – 4.0
Fort Worth – 3.7
Flexible Packaging – 3.7
Virtual Color – 3.6

The scores and comments were reviewed with each General Manager for improving associate engagement score.

When we look back, we find, our corporate staff, division managers and associates have accomplished a lot; but, we need to do more. We need to continually update our safety manual, select the safe division of the year and increase associate awareness to plant hazards and necessary safety precautions. This increase in awareness will assure that each one of us looks out for each other to prevent an accident, so it never has a chance to happen since our safety goal is:

* to have zero safety violations at all divisions
* achieve zero lost time accidents
* increase safety awareness score of at least 90 percent
* achieve OSHA VPP recognition, and
* achieve the best safety record in the printing industry

SAFETY IS A TEAM EFFORT

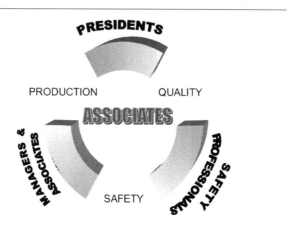

Our CEO, Rich Adler, Jr., summarized our achievement as, "What sets apart the extraordinary results from the good results is that safe working practices have become interwoven within the culture of those (our) facilities. Safe practices are how they get the job done rather than something they do in addition to their job." We have made significant progress, but we are not satisfied. We ask our associates to remember, together we can achieve what we dream; and our dream is to have the best safety record in the printing industry!

Labeling Fort Dearborn
a Safe Company

When a printing company stops printing to install a ventilation system, you know safety is a way of doing business.

BY SANDY SMITH

Fort Dearborn Co. of Niles, Ill., employs approximately 900 workers at six U.S. facilities and one in Mexico. The company prints labels that go on paint cans, ketchup bottles and soft drink bottles.

While installing a new, six-color printing press at one of Fort Dearborn Co.'s facilities, it was determined that the ink mixing room had to be moved to a new area. The new room did not have exhaust ventilation. Ink was necessary to continue the production, but proper ventilation was necessary to continue making ink in a safe manner. Production stopped until ventilation could be installed.

"There was no question about continuing production," says Michael Saujani, corporate safety director at Fort Dearborn. "We could not put our employees at risk."

Employees respond well to management's commitment to safety, says Saujani. "They see management emphasis on safety is not just talk. When we first started, employees would bring up issues about safety and it would take management a couple of months to do something about it. Now, it gets done in a day or two. Employees see results, and they respond to it. If their concerns are not corrected right away, you get feedback from them about it."

In addition to quick response to employee concerns, the company pays for safety gear required by associates, has conducted safety fairs for the past three years and has provided free onsite medical screening for associates for blood pressure, glucose levels, body fat and cholesterol.

"Healthy employees are productive employees," says Saujani. "That's good business."

Focus on Ergonomics

Because of an aging work force, Saujani invited a professor from Northern Illinois University's graduate program in ergonomics to send his students out to Fort Dearborn's facilities and analyze ergonomics. The review, which cost the company nothing since it was incorporate into their semester studies, focused on material handling issues, since about 70 percent of the company's ergonomic injuries were related to sorting and palletizing.

The suggestions from the ergonomics students included buying vacuum lifts and lift assists to reduce bending. The company purchased new equipment and, at the same time, trained managers to identify ergonomic hazards before they become an issue and employees were hurt.

"For example, a customer wanted us to pack labels in a box and the box weighed 60 pounds. We normally would not lift anything above 35 pounds. The manager stepped in and told the customer, 'We can't do this because we went through ergo training and we know it's not safe for our associates to lift 60-pound boxes.'"

The customer agreed to change its requirements from 60-pound boxes to 35-pound boxes. The result, says Saujani, are boxes that are easier to lift, thereby reducing the potential for injury not only for

Fort Dearborn employees, but for anyone who has to move those boxes farther down the line.

In addition to equipment and training, the company invited a chiropractor to one of its safety fairs. She conducted classes every 30 minutes that taught employees how to stretch and gave them instruction about proper posture and lifting techniques. The company is thinking about requiring employees to stretch for 5 minutes before the start of each shift, but has not instituted that program yet, says Saujani.

The company's focus on hazard prevention and injury reduction has paid off, he adds. "I met an insurance broker and he said the company had saved $1 million [in injury and illness-related costs] in three years. That is substantial. Safety does make good business sense. But the most important thing is you send people back to their families safe every day."

Last to First: How to Transform a Failing Safety Program

Picture your company on OSHA's hit list of high hazard companies targeted for inspection, your workers' compensation premiums have skyrocketed 425 percent in four years, and your lost-workday injury rate is 7.63 while the industry average is 2.7. You've just been hired for the job of safety manager and told to fix the problems. What do you do?

Michael Saujani, CSP, was put in this position at a large manufacturing company with several plant locations in the United States several years ago, and with the company's support managed to turn around its safety performance to now have one of the best safety records in its industry, reduced its workers' compensation experience modification rate from 1.5 to 0.9, and has a lost-workday rate of 1.31 company-wide and 0.61 at one of the facilities.

Mr. Saujani talked about this turnaround at a conference of more than 600 safety professionals sponsored by the Indiana chapter of the American Society of Safety Engineers (ASSE) and the Indiana Chamber of Commerce in partnership with the National Environmental, Health and Safety Training Association (NESHTA), Indiana INSafe, and the American Industrial Health Association.

Sad State of Affairs

Here is the state of affairs that Mr. Saujani started with on day one of his new assignment as Safety Manager:

- No monthly Safety Committee meetings at all locations.
- Volunteers were serving as Safety Team Leaders at each facility.
- No professional safety director.
- Workers and managers were not fully recognized for their safety efforts.
- All department managers were not fully involved in safety.
- All division General Managers were not fully involved in safety.
- Safety culture was inconsistent.
- Safety responsibilities were delegated to plant engineering.
- Hazards were not being corrected immediately.
- Worker training was inconsistent.
- Records of training and inspections were not available.

Road to Success

Here is a summary of the process that Mr. Saujani developed and followed to turn things around:

1. Develop a road map
2. Define roles and responsibilities for workers and management
3. Develop a safety program
4. Monitor safety performance
5. Get senior management involved in this entire process

1. Develop a road map

He looked at the big picture of where he was and wanted to go, and applied this roadmap to each phase of the 5-step process summarized above:

- *Structure* – The structure of the company was analyzed in related to safety, including the corporate safety team, division safety teams, insurance program, safety committees, and the written safety program.
- *Processes* – Processes and procedures related to safety performance, including specific safety efforts (e.g., back injury prevention, chemical exposure prevention, etc.), an analysis of what was wrong with each safety effort, the return-to-work program, accident investigation, job hazard analysis, audits, certifications, safety awareness projects, and safety incentives.
- *People* – Review and enhance the roles of the CEO, safety director, division leadership, general plant managers, floor managers and shift supervisors, and "associates."
- *Culture* – Review current attitudes and practices, then engage associates in the process, promote safety awareness, encourage safe behavior, create hazard-free work areas, and achieve recognition as an industry leader in safety.

2. Define roles and responsibilities for workers and management

He identified and redefined the roles and responsibilities of each layer of management. For example, he persuaded the CEO to adopt the following goal as his responsibility:

"The role of the CEO is to furnish each associate a place of employment which is free from recognized hazards that could cause injury of any type, serious or minor, to our associates and visitors."

He formally defined the responsibilities of the Safety Director, Safety Team leaders, department and plant managers, supervisors, and associates. For example, the role of each associate is to:

- Obey all safety rules and regulations
- Participate in the training meetings
- Make suggestions on how to improve the system
- Work safely as an individual and as a team
- Ensure work area is kept clean and free of hazards
- Report all unsafe acts and conditions to the immediate supervisor

3. Develop a safety program

The safety program was developed using the roadmap as a guide: he looked at company structure, processes, people, and culture. It wasn't just development of written safety procedures, it included incentives for management at all levels to buy in to the safety program. For example, the company moved to a high-deductible workers' compensation insurance system. If one facility incurs higher workers' compensation costs than other facilities within the company, the level of bonuses to staff would be adjusted downward according to a formula tied to the workers' compensation rate for the whole company. Also, all managers and supervisors are required to obtain an OSHA 10-hour training program certificate, on company time.

Mr. Saujani also used control charts as metrics to measure performance using facility statistics for lost-workdays and other indicators of safety activity. Control charts are spreadsheet bar or line charts that show trends in safety performance.

He discovered during the job hazard analysis that mechanical devices would reduce the high prevalence of back and shoulder injuries during packaging and shipping operations. Once the devices were put in place, he measured the productivity of associates performing their work. He found that associated increased the rate of completion of tasks by 20 percent and eliminated lost-workday injuries related to back and shoulder.

4. Monitor safety performance

Safety performance was not just measured by fewer injuries. It also monitored the efforts of associates to do the right thing so such efforts could be recognized and rewarded.

5. Get senior management involved

Senior management was initially motivated to improve safety performance with the consequences of a failing safety program and a desire to correct mistakes. The motivation was sustained once the roles and responsibilities of management were well defined and safety goals were created that are tied to management accountability.

Goal of Continuous Improvement

Mr. Saujani's organization has established specific goals that require continuous improvement in safety performance:

- Zero safety violations at all plants
- Achieve zero lost time accidents
- Increase safety awareness score to 90 percent
- Achieve OSHA VPP status
- Achieve best safety record in the industry

He believes these goals will be accomplished through a team effort at all levels of the organization.

For more information about Mr. Saujani's safety program, contact him at mikesaujani@gmail.com.

About the Author

Photo by Andrea Ferenchik

Michael Saujani, CSP, CPCU, ALCM, ARM, is the president of MKS Safety, LLC, providing loss-control consulting services to various clients in the Chicago area.

He was a corporate safety director for Fort Dearborn Company from 2002 to 2013. During his leadership at FDC, he was able to put together a road map and safety structure that changed the safety culture at the company to the extent that EHS (Occupational Hazards) recognized the company as one of the twelve best for the year in 2005 and CNA Insurance Company recognized the company with a Leadership in Safety award in 2012.

Prior to his tenure at FDC, he worked in the loss-control department for Amerisure Mutual Insurance, Hanover Insurance, Hartford Insurance, and Allstate Insurance both as a senior loss-control consultant and supervisor offering loss-control services to insurance company clients and mentoring several field loss-control consultants.

Michael holds a bachelor's degree in mechanical engineering and graduated with first class honors. He is a professional member of the American Society of Safety Engineers and a member of the Certified Property and Casualty Underwriters Society (CPCU). As a professional member of ASSE, he won a NEIL Chapter Safety Professional of the Year (SPY) award in 2012 and ASSE Region V SPY in 2015. He was ASSE Northeastern Illinois chapter president for the year 2014-2015. The NEIL chapter was recognized with a Platinum Award for the first time that year.

He has written several safety and risk management articles that have been published in professional safety and insurance risk management journals.

Michael and his wife Meena raised two very successful daughters, a doctor and a lawyer.

Made in the USA
Middletown, DE
11 March 2017